A KEY TO THE ADULT BRITISH WATER BEETLES

LAURIE E. FRIDAY
Department of Applied Biology, Pembroke Street, Cambridge CB2 3DX
Research Fellow, Darwin College, Cambridge

CONTENTS

INTRODUCTION	1
EXAMINING BEETLES AND SOME CHARACTERS USED IN THE KEY	2
USING THE KEY	6
COLLECTING WATER BEETLES	7
REFERENCES AND FURTHER HELP	8
ACKNOWLEDGEMENTS	9
INDEX TO KEYS	10
KEYS	12
TABLE 1. COLOUR GUIDE TO COLYMBETINAE AND DYTISCINAE	142
TABLE 2. COLOUR GUIDE TO HYDROPORINAE	144
TABLE 3. SIZES OF WATER BEETLE GENERA	146
SPECIES CHECKLIST, DISTRIBUTIONS AND HABITAT NOTES	147

INTRODUCTION

Of the very large number of species of Coleoptera found in Britain, relatively few occur in aquatic habitats. As far as is known, no British beetle spends its entire life-cycle in water, since the pupal stage usually occurs on land. However, more than 250 species of Coleoptera, drawn from a wide range of families, spend part of their life history in an aquatic environment, to which they show varying degrees of adaptation.

This key incorporates members of 14 families, belonging to 2 sub-orders:

> ADEPHAGA   Gyrinidae; Haliplidae; Hygrobiidae; Noterridae;
> (=Hydradephaga)   Dytiscidae.
> POLYPHAGA   Hydrophilidae; Hydraenidae; Dryopidae; Elmidae
>   (=Elminthidae); Scirtidae (=Helodidae); Heteroceridae;
>   Limnichidae; Chrysomelidae (sub-family Donaciinae);
>   Curculionidae.

The Adephaga spend the egg, larval and adult phases in water. The adults are primarily carnivorous, and show marked adaptations for swimming; most have a streamlined body shape, and, in the Dytiscidae, the hind legs are long and slightly flattened, with fringes of long hairs. In all five families, the hind coxae (see Fig. 3) are immovable and restrict the leg movement to one plane, which improves their swimming efficiency. The Gyrinidae ("whirligig beetles") are unique in having short, paddle-like legs and horizontally divided eyes, which function as efficiently on the water surface as below it.

The aquatic Polyphaga is a diverse group, with a variety of life-styles and adaptations. Many of the Hydrophilidae are truly aquatic (some have streamlined bodies and swimming legs), others are fully terrestrial, whilst another group lives in detritus around the water margin. The Dryopidae and Elmidae ("riffle beetles") are also aquatic in all but the pupal stage, but the Hydraenidae are aquatic only as adults. The adults of all these families are generally poorly adapted for swimming and tend to walk along the bottom. Of the remaining families, the Scirtidae and most of the Donaciinae (Chrysomelidae) are aquatic as larvae. The adults live in vegetation around the water's edge. A number of weevil species (Curculionidae) are associated with aquatic food plants. A few members of the Heteroceridae and Limnichidae are found in the banks of freshwater and brackish habitats. Members of these last five families may turn up in net samples so they are included here, though not keyed out to species.

In view of the taxonomic diversity encompassed by the term "water beetles", it is not surprising that representatives occur in a very wide range of aquatic and semi-aquatic habitats, from damp moss, peaty pools, streams and canals to coastal rocky shores and salt marshes. The species checklist (p. 147) gives information on the habitats and the present state of knowledge of the species' distribution within the British Isles.

BEETLE STRUCTURE

It is important to distinguish adult water beetles from water mites (Hydracarina) and true bugs (Hemiptera). Beetles have six legs (mites have eight) and their hardened forewings meet in a straight line down the centre of the back (Fig. 1a). The wings overlap at the rear in adult bugs, e.g. water boatmen (Fig. 1b).

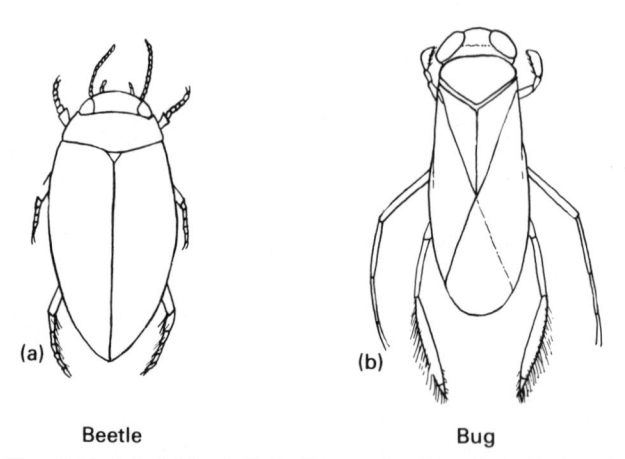

Beetle   Bug

Figure 1. Typical adult beetle (Order Coleoptera) and bug (Order Hemiptera).

This key uses a simplified terminology for structures, shapes and textures; the terms used for parts of the beetle are shown in Fig. 2, the upper side of a dytiscid, and Figs 3 and 4, the undersides of a dytiscid (Adephaga) and a hydrophilid (Polyphaga) respectively.

EXAMINING BEETLES AND SOME CHARACTERS USED IN THE KEY

Whenever possible, and particularly in the keys to families and genera, "macroscopic" characters visible at a magnification of × 10 are used. When separating species, however,

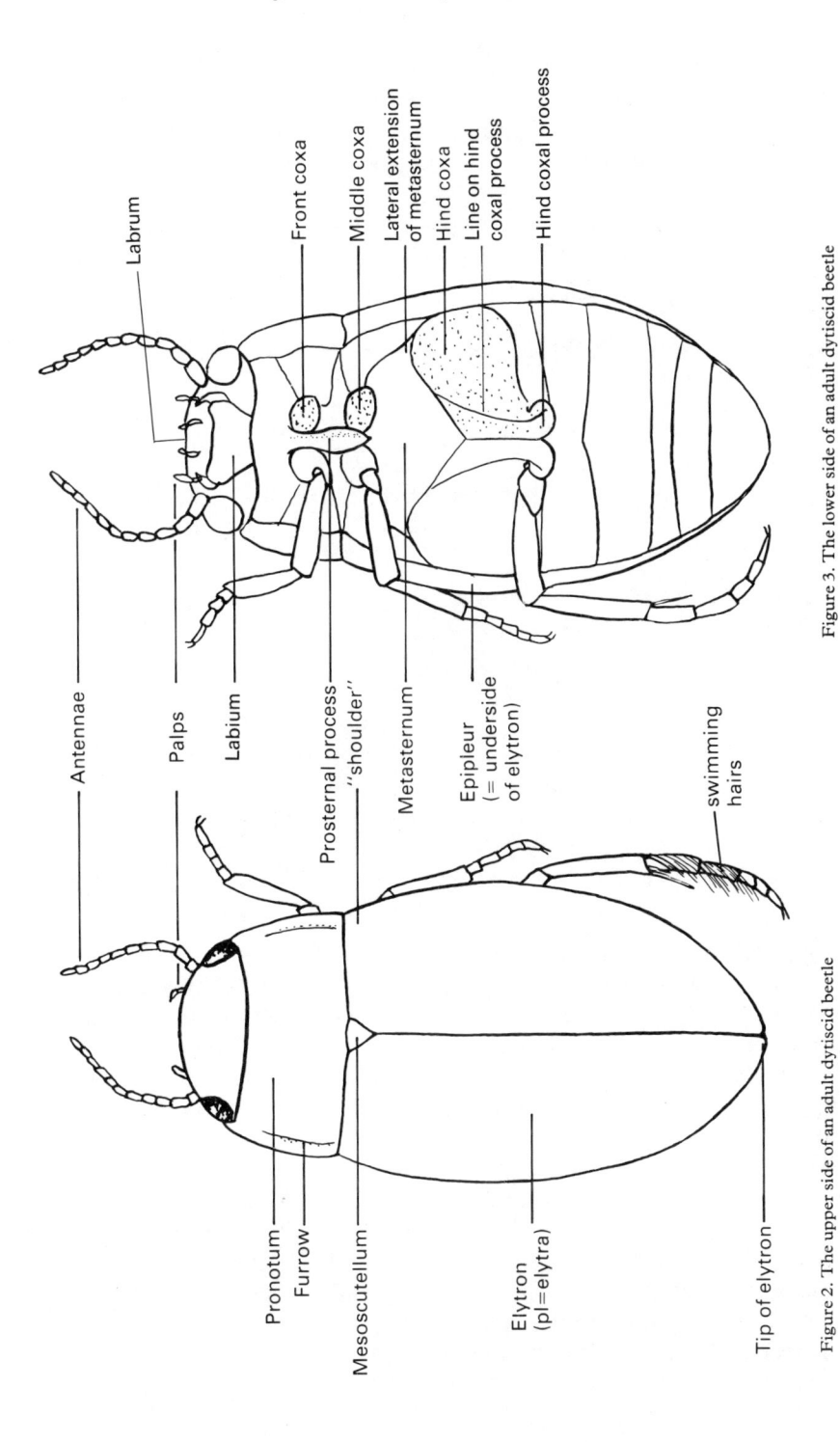

Figure 3. The lower side of an adult dytiscid beetle

Figure 2. The upper side of an adult dytiscid beetle

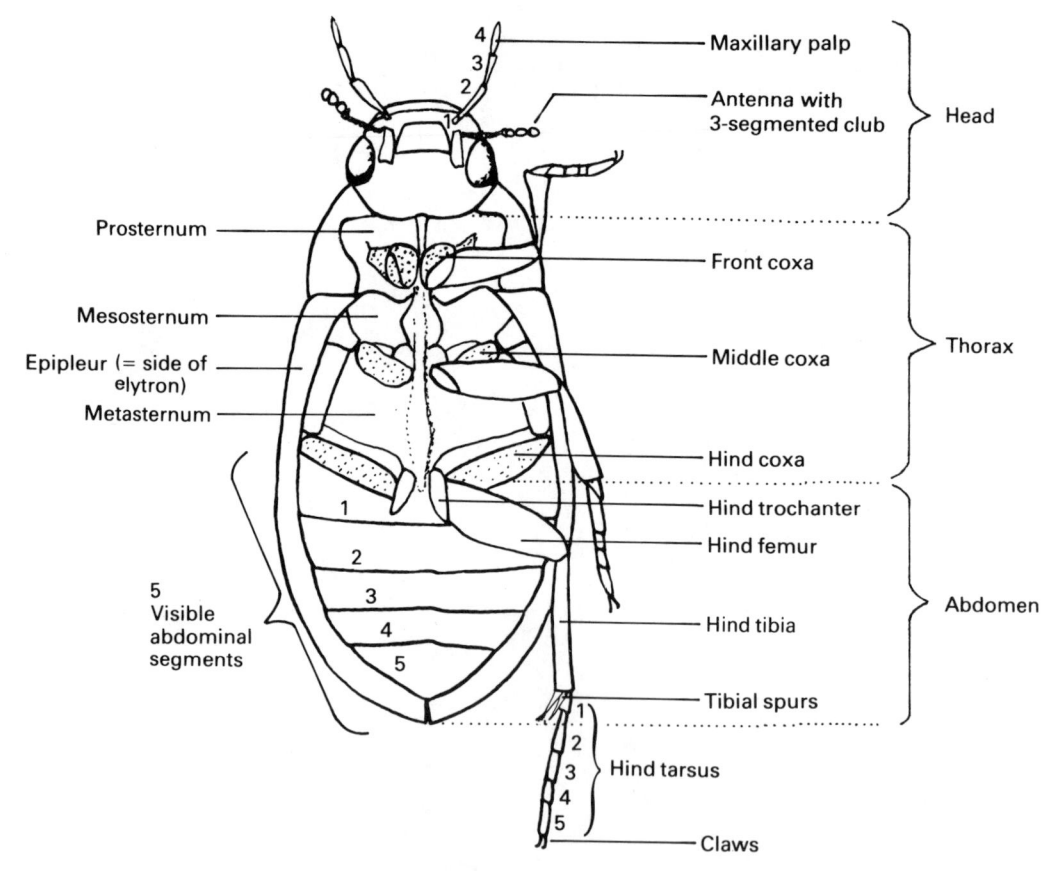

Figure 4. The underside of an adult hydrophilid beetle.

characters requiring higher magnification are often necessary; the minimum magnification needed to see these characters is given in the text (*e.g.* ×25). In some genera of small beetles, a binocular microscope magnifying to at least ×30 and up to ×80 may be necessary. In such cases, the determination of species may be beyond the scope of those without access to a suitable microscope. Good lighting is also essential for many of the characters; a small directable light source, preferably with a frosted bulb, is better than a diffuse source.

Water beetles, by virtue of their shape, may be difficult to manipulate. Live specimens can be temporarily immobilised if transferred to a dry tube and given a whiff of CO_2 administered through the open neck of the tube with a CO_2-type bottle opener. Dead specimens can be manipulated under alcohol or water, but must be viewed DRY at those points of the key indicated. If necessary, the specimen can be steadied in a groove in a small blob of Blu-tak. If pinned (see next section), specimens can readily be turned to be examined from any direction. The key uses many characters on the underside and card-mounted specimens will need to be gently floated off the card by application of a little water on a soft watercolour brush.

Completely dried-out specimens can be made pliable, and so examined without fragmenting, by soaking for an hour or so in hot (not boiling) water.

Sizes: are measured from the front of the head to the tip of the elytra (except in the Elmidae and Dryopidae, where the head may be retracted and the front of the pronotum is used instead). N.B. Specimens which have been preserved in alcohol or left to soak in water will often have the head protruding and the abdomen extended beyond their normal positions.

Colours: are used where colour patterns are characteristic. Users should note, however, that geographical, habitat and genetic variations occur (the key allows for this) and specimens may not match illustrations *exactly*. Also, colours tend to fade in dry collections and are darkened in alcohol; in the latter case, specimens should be allowed to dry for several minutes, at least until the spirit evaporates from underneath the elytra, before the colours develop.

External texture and structures: are only apparent when the beetle is DRY, is CORRECTLY ORIENTATED and ILLUMINATED, and is viewed at SUFFICIENTLY HIGH MAGNIFICATION. Instructions are given in the key, as necessary. N.B. When viewing specimens to be preserved in spirit, do not let the internal organs dry out completely; only the outside need be dry.

Antennae and palps: are useful for distinguishing the major families of water beetles (see Family Key). It is important to distinguish between these two types of head appendage:

palps are mouthparts and appear to have only 3 segments and never have a hairy club at the end;

antennae have many more segments, usually 7–11 and may or may not be clubbed. In Hydrophilidae, Hydraenidae, Elmidae and Dryopidae, especially in live specimens, the antennae and palps are normally hidden in pockets on the underside of the head (Figure 5) and will need to be teased out gently for examination.

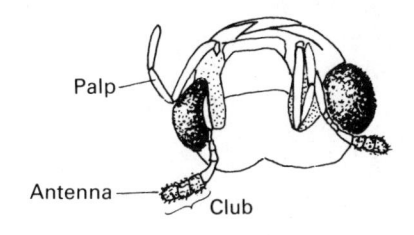

Figure 5. The underside of the head of an adult hydrophilid beetle

Genitalia and secondary sexual characters. In many genera, males and females can be distinguished by external characters, which may themselves be useful for recognising species. For example, in the Colymbetinae and Dytiscinae, the males alone have widely expanded front tarsal segments bearing pads of sucker hairs, while in the latter group, only the females may have longitudinal grooves in the elytra; in the Haliplidae and the majority of Hydroporinae, males have expanded front and middle tarsal segments and may have modified (longer, more abruptly curved, unequally broadened and/or toothed) front tarsal claws (these distinctions are not always obvious in *Hydroporus* and both sexes have sucker hairs on the front and middle tarsi); some Hydrophilidae also have modified male front tarsi and, in some genera of Hydraenidae, males appear to have one less sclerotised abdominal segment than the females. Males and females of many species are of slightly different sizes or shapes, and in many dytiscid species, females may be found which appear duller than the shiny males.

In several genera, however, the genitalia are the best clue to the sex of the specimen and may be the ONLY sure means of identifying to which species it belongs. The keys to *Gyrinus, Haliplus, Hydroporus, Oulimnius, Helophorus* and *Enochrus* necessarily rely on characters of genitalia, particularly of the males.

Genitalia can be extruded from some fresh specimens by gentle pressure on the abdomen (some live beetles, particularly *Gyrinus,* will extrude the tips sufficiently for identification without persuasion). In most cases, however, it will be necessary to draw out, or dissect out, the genitalia.

Fresh and soft spirit-preserved beetles may be secured on their back, held with forceps if necessary, and fine forceps inserted between the dorsal and ventral plates at the tip of the abdomen to gently pull out the genitalia. Alternatively, if the elytra separate easily, the genitalia may be teased out via the softer dorsal surface of the abdomen between the tips of the slightly displaced elytra.

In some cases, however, it will be necessary to remove the abdomen. Dissection is likely to destroy a number of important characters useful for identification, and these should be examined before removing the genitalia. Dried-out specimens will need first to be softened in hot water. The beetle should be steadied on its back and the last few segments of the abdomen removed using fine mounted needles or a scalpel. Turning the tip of the abdomen over, the genitalia may be extracted through the softer dorsal surface (further softening in hot water may be required). The genitalia are sclerotised and lie within softer tissues which can be gently picked off in a drop of water using fine mounted needles. For a carded collection (see below), they may be mounted next to their owner in a spot of gum or D.M.H.F. (dimethyl hydantoin formaldehyde made up in water or 70% alcohol—see below for suppliers) but will need to be moistened for subsequent examination.

The genitalia must be displayed in the correct orientation, as described in the keys, and viewed, moistened and well-lit, at very high magnification ($\times 50$ and above if possible). Where greater resolution is required, as with small species and difficult genera, the male capsule should be mounted in a suitable mountant, such as Canada balsam, under a coverslip on a microscope slide and viewed under a compound microscope. NB. The pressure of the coverslip may tend to alter the relative positions and orientations of the constituent parts; there are warnings about this in the key.

Figure 6 shows the range of types of genitalia found in water beetles. Generally, males have three main parts at the tip: a central strongly sclerotised *penis* or *aedeagus* and two lobes, *parameres,* one on each side. The two parameres may be similar (*e.g.* Colymbetinae) or dissimilar (Haliplidae). The whole male apparatus may take the form of a capsule with a unifying basal part (*e.g.* Hydrophilidae). Females usually have only two main parts at the tip, appearing as flaps, lobes or long fingers, and the ovipositor is inconspicuous. However, where eggs are laid inside plant materials (*e.g.* Laccophilinae), there is a well-developed central ovipositor.

Using the Key

Always start at Couplet One in the Family key (p. 12). Follow the trail to subsequent couplets as shown at the end of the correct half. You will be led to your end point via a series of sub-family, genus and species keys, as shown in the Index to Keys (p. 10). Use the general descriptions and size ranges at the head of each key to check that you are on the

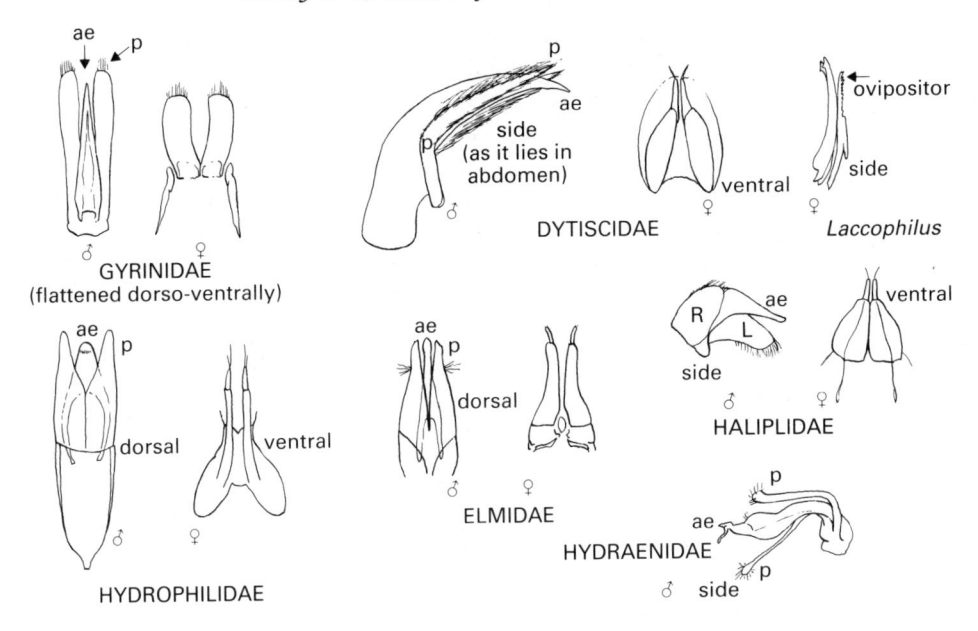

Figure 6. Range of genitalia found in some water beetle families (ae = aedeagus, p = paramere, L = left, R = right)

right track. Identifications are taken to species wherever possible (except for Scirtidae, Dryopidae, Heteroceridae, Chrysomelidae and Curculionidae).

Illustrations: appear opposite the relevant half of the couplet and are referred to within that couplet by the letters (a), (b), etc.

Tables 1, 2 and 3 (pp. 142–146): are additional aids to identification based on the size ranges of genera in the main aquatic families and on the colour patterns in Dytiscinae, Colymbetinae and Hydroporinae; they should be used only in conjunction with the keys.

Nomenclature mainly follows the Royal Entomological Society checklist (Pope, 1977) and the Checklist (p. 147) gives references for the species new to Britain since 1977. Synonyms found in such standard works as Balfour-Browne (1940, 1950, 1958) and Joy (1932) are given in the Species Checklist at the end of the key. This checklist also gives an indication of the habitats, distributions and abundance of individual species.

COLLECTING WATER BEETLES

Useful details of equipment and techniques are given by Balfour-Browne (1940) and in the Balfour-Browne Club (B-B.C.) Newsletters 10, 13 and 17 (see references); only a brief outline will be given here.

Methods of collecting water beetles are as many and varied as are the beetles and the habitats they occur in. A net consisting of a mesh bag on a strong double metal frame is extremely useful amongst vegetation and in streams. Searching and catching with the fingers is effective for slow- or non-swimming beetles, while a teaspoon might catch the smallest specimens. Simple baited funnel traps have proved useful for catching larger carnivorous beetles, such as Dytiscidae.

Although many species of water beetle do occur in open water in ponds, lakes, rivers and streams, many live in damp vegetation: *Sphagnum* moss is a particularly rewarding habitat for water beetle collecting. The damp vegetation can either be put in a dish and sorted by hand, or a *small* area can be stamped on and a net used in the resulting puddles.

Samples are best sorted on a white polythene sheet or in a white tray. Beetles can be transported home alive in water or damp moss in small specimen tubes, or killed with 70% alcohol or ethyl acetate vapour. N.B. Ethyl acetate attacks plastic tubes.

To set up a card-mounted reference collection, beetles should be arranged and glued onto individual cards as soon as possible after death, and cleaned gently, using a fine paintbrush dipped in ethyl acetate. Water-soluble glues, such as gum tragacanth, are recommended. It is essential that each specimen should be accompanied by a label, on the same pin, giving the date of capture, locality (with grid square), type of habitat, and, if possible, the species name.

The most useful reference collections have several specimens of each species, both males and females, with one or two mounted on their backs.

For several difficult genera (e.g. *Gyrinus, Hydroporus, Haliplus,*) it is advisable to extract the genitalia (see previous section) as soon as practicable and mount them in a drop of gum or D.M.H.F. next to the beetle.

An alternative way to keep a dry collection is to pin the beetle with an appropriately sized entomological pin through the shoulder area of one elytron, taking care to avoid the midline on the underside and thus not to destroy any characters required for identification.

Dry collections should be kept in dust-proof boxes, lined with cork or polyurethane foam (NOT expanded polystyrene as this reacts disastrously with ethyl acetate and some other chemicals and does not grip pins securely) and protected with a small piece of naphthalene.

Collections kept in small bottles of 70% alcohol are also useful, allowing all aspects of the specimen to be easily seen (although the caveat about colours in the previous section should be noted) and take up very little space. Labels written in pencil can be inserted into the tubes with the specimens.

Entomological equipment, such as pins, collection boxes, killing and relaxing fluids and nets, is available by mail order from a number of specialist suppliers, such as Watkins & Doncaster. D.M.H.F. can be obtained from The Chemical Intermediates Co. Ltd., Barnfields Industrial Estate, Leek, Staffs. Sound advice on the choice of a pond net can be found in the Balfour-Browne Club Newsletter *13*.

References and Further Help

Anyone interested in water beetles should consider joining the Balfour-Browne Club. The Club was started in 1976 and membership is open to anyone with an interest in water beetles. There are about 200 members in twenty countries and one of the main objectives of the Club is to further the study of water beetles on a European basis.

Beginners are specially welcome with a subscription discount for students and an identification service run in conjunction with the British and Irish recording schemes.

The Club produces three or four newsletters a year and these include records, paper reviews and keys to the more difficult groups. There is usually at least one field meeting a year.

The secretary is: Garth Foster, 20 Angus Avenue, Prestwick, Ayrshire KA9 2HZ, Scotland.

The following books are invaluable:

Balfour-Browne, F. (1940, 1950, 1958). *British Water Beetles*. Vols. 1, 2, 3. Ray Society.
Hansen, M. (1987). *The Hydrophiloidea (Coleoptera) of Fennoscandia and Denmark*. Fauna Entomologica Scandinavica **18**. E. J. Brill/Scandinavian Science Press Ltd., Leiden/Copenhagen.
Joy, N. H. (1932). *A Practical Handbook of British Beetles*. Reprinted (1976) in reduced format by E. W. Classey Ltd., Park Road, Faringdon, Oxon SN7 7QR.
Oldroyd, H. (1970). *Collecting, Preserving and Studying Insects*. 2nd edition. Hutchinson & Co. Ltd., London.

These papers are referred to in the text:

Angus, R. B. (1978). The British species of *Helophorus*. *Balfour-Browne Club Newsletter* **11**.
Angus, R. B. & Carr, R. (1982). *Gyrinus natator* (L.) and *G. substriatus* Stephens as British species. *Entomologist's Gazette* **33**: 223–229.
Berge Henegouwen, A. van (1986). Revision of the European species of *Anacaena* Thomson (Coleoptera: Hydrophilidae). *Entomologica Scandinavica* **17**: 393–407.
Berthélemy, C. (1979). Elmidae de la région Paléarctique Occidentale: systématique et répartition (Coleoptera, Dryopoidea). *Anales de Limnologie* **15**(1): 1–102.
Carr, R. (1984a). *Limnebius crinifer* Rey new to Britain, with a revised key to the British *Limnebius* species (Coleoptera: Hydraenidae). *Entomologist's Gazette* **35**: 99–102.
Carr, R. (1984b). A *Coelambus* species new to Britain (Coleoptera: Dytiscidae). *Entomologist's Gazette* **35**: 181–184.
Clarke, R. O. S. (1973). Coleoptera, Heteroceridae. Handbooks for the Identification of British Insects 5(2c). Royal Entomological Society of London.
Foster, G. N. (1979). *Hydraena* aedeagophores. *Balfour-Browne Club Newsletter* **13**.
Foster, G. N. (1981). Figures of Gyrinidae. *Balfour-Browne Club Newsletter* **20**.
Foster, G. N. (1984). Notes on *Enochrus* subgenus *Methydrus* Rey (Coleoptera: Hydrophilidae), including a species new to Britain. *Entomologist's Gazette* **35**: 25–29.
Foster, G. N. & Angus, R. B. (1985). A Key to British *Hydroporus*. *Balfour-Browne Club Newsletter* **33**.
Foster, G. N. & Phillips, E. J. (1983). *Laccobius simulator* D'Orchymont (Coleoptera: Hydrophilidae) confirmed as British. *Entomologist's Gazette* **34**: 265–266.
Foster, G. N. & Spirit, M. (1986). *Oreodytes alpinus* new to Britain. *Balfour-Browne Club Newsletter* **36**.
Freude, H., Harde, K. W. & Lohse, G. A. (1971). eds. *Die Käfer Mitteleuropas* 3. Goecke & Evers. Krefeld.
Gentili, E. (1977). The British species of *Laccobius*. *Balfour-Browne Club Newsletter* **4**: 8–10.
Olmi, M. (1978). *Driopidi, Elmintidi*. Guide per il riconoscimento delle specie animali delle acque interne Italiane 2. Consiglio nazionale delle ricerche, Verona.
Parry, J. (1980). Comparison of *Oulimnius* genitalia. *Balfour-Browne Club Newsletter* **18**.
Parry, J. (1983). *Haliplus varius* Nicolai (Col. Haliplidae) new to Britain. *Entomologist's Monthly Magazine* **119**: 13–16.
Pirisinu, Q. (1981). *Palpicorni*. Guide per il riconoscimento delle specie animali delle acque interne Italiane 13. Consiglio nazionale delle ricerche, Verona.
Pope, R. D. (1977). *A checklist of British Insects 3: Coleoptera and Strepsiptera*. Handbooks for the Identification of British Insects 9(3). Royal Entomological Society of London.
Unwin, D. M. (1984). A key to families of British Coleoptera and Strepsiptera. *Field Studies* **6**(4): 149–197.

The following contains a key to the genera of the larvae of most of the families of water beetles included in this key:

Richoux, P. (1982). *Coléoptères Aquatiques*. Bulletin mensuel de la Société Linnéenne de Lyon 51e année no 4, 8+9 (extaits). Association Française de Limnologie. (Dr R. Ginet, Université Claude-Bernard, Lyon 1, Départment de Biologie Animale et Ecologie, 43 Bvd du 11-Novembre 1918, 69622 Villeurbanne, France.)

Acknowledgements

Much of the work of preparing this key was carried out in the Insect Room of the Museum of Zoology, Cambridge, and I should like to thank Professor G. Horn and Dr W. A. Foster for allowing me access to the collections and providing bench space.

I am most grateful to many people who gave expert advice, especially Drs R. B. Angus and P. Langton and Messrs R. Carr and D. M. Unwin. I owe a special debt of gratitude to Dr G. N. Foster for his unstintingly generous help and for initially firing my enthusiasm for water beetles. Drs A. E. Friday and S. A. Corbet have provided a good deal of encouragement during the production of this key.

Many of the figures in the text have been drawn by Mrs Marilyn Crothers, and I am pleased to acknowledge her skilful and patient contribution. The beautiful painting on the front cover is the work of Sarah Wroot and has been provided by the Balfour-Browne Club.

Finally, I should like to thank over 100 anonymous testers who sent written reports on the Test Version, resulting in substantial and, I hope, beneficial modification of the text, and especially Dr S. Tilling, who has co-ordinated this project with considerable skill and forebearance.

I hope that this Key will encourage more people to study water beetles; it is with this aim in mind that I respectfully dedicate the Key to the members of the Balfour-Browne Club, present and future.

Editor's Acknowledgement

The Field Studies Council is grateful to the Linnean Society of London for a donation from the Omer-Cooper Bequest towards the cost of publishing this key.

Index to Keys

FAMILY KEY p.12.
HYGROBIIDAE, SCIRTIDAE (HELODIDAE), LIMNICHIDAE, HETEROCERIDAE, CURCULIONIDAE

A GYRINIDAE, p. 18
   Orectochilus, Gyrinus 11 spp

B HALIPLIDAE, p. 22
   Brychius, Peltodytes
     B1 *Haliplus* 17 spp, p. 22

C NOTERIDAE, p. 30
   Noterus 2 spp.

D DYTISCIDAE p. 31
   D1 Colymbetinae, p. 33
     Colymbetes, Copelatus, Platambus
       D1a *Rhantus* 6 spp, p. 35
       D1b *Ilybius* 7 spp, *Agabus* 20 spp, p. 38
   D2 Dytiscinae, p. 50
       D2a *Dytiscus* 6 spp, p. 51
       D2b *Hydaticus* 2 spp, p. 53
       D2c *Acilius* 2 spp, p. 53
       D2d *Graphoderus* 3 spp, p. 54
   D3 Laccophilinae, p. 55
     Laccophilus 3 spp
   D4 Hydroporinae, p. 56
     Hydrovatus, Stictotarsus, Laccornis, Deronectes, Scarodytes, Porhydrus

      D4a *Hyphydrus* 2 spp, p. 62
      D4b *Hygrotus* and *Coelambus* 4 + 5 spp respectively, p. 63
      D4c *Bidessus* and *Hydroglyphus* 2 + 1 spp respectively, p. 66
      D4d *Graptodytes* and *Stictonectes* 4 + 1 spp respectively, p. 67
      D4e *Oreodytes* and *Scarodytes* 4 + 1 spp respectively, p. 68
      D4f *Hydroporus* and *Suphrodytes* 28 + 1 spp respectively, p. 70
      D4g *Potamonectes* 4 spp, p. 84

E CHRYSOMELIDAE, p. 86
  Macroplea, Plateumaris, Donacia, Halticinae, Galerucinae

F ELMIDAE, p. 88
  Macronychus, Stenelmis, Elmis, Limnius, Esolus
      F1 *Riolus* and *Normandia* 2 + 1 spp respectively, p. 90
      F2 *Oulimnius* 4 spp, p. 91

G HYDROPHILIDAE & HYDRAENIDAE, p. 93
  Spercheus, Georissus
  Coelostoma, Megasternum, Cryptopleurum, Sphaeridium
  Hydrophilus, Hydrochara, Limnoxenus,
  Hydrobius, Chaetarthria, Cymbiodyta
  Helophorinae, p. 101
      G1 *Helophorus* 17 aquatic spp, p. 101
  Hydrochinae, p. 111
      G2 *Hydrochus* 6 spp, p. 111
  Sphaeridiinae, p. 113
      G3 *Cercyon* 9 aquatic spp, p. 113
  Hydrophilinae
      G4 *Berosus* 4 spp, p. 117
      G5 *Helochares* 3 spp, p. 119
      G6 *Enochrus* 10 spp, p. 120
      G7 *Laccobius* 9 spp, p. 123
      G8 *Paracymus* and *Anacaena* 2 + 4 spp respectively, p. 127
  HYDRAENIDAE
      G9 *Limnebius* 5 spp, p. 129
      G10 *Hydraena* 10 spp, p. 131
      G11 *Ochthebius* 15 spp, p. 135

H DRYOPIDAE, p. 141
  Dryops 7 spp, *Helichus*

FAMILY KEY

There is no single character, or set of characters, for recognising an aquatic beetle; most beetles likely to occur in net samples are keyed out below. If your specimen will not key out here, it is probably terrestrial and has fallen in—see Unwin (1984).

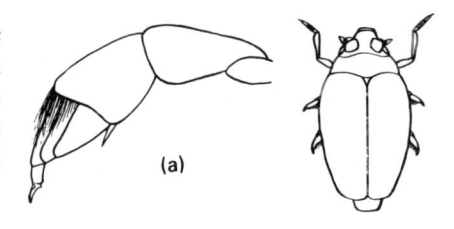

1   Middle and hind legs shorter than the front legs and very broad (a); eyes divided horizontally, giving the appearance of 4 eyes; swim on the surface of the water; size range 3.5–7.8 mm . . . **GYRINIDAE** Key A, p. 18

—   Middle and hind legs as long as the front legs; eyes undivided; size range 1.0–48.0 mm 2

2   Antennae long and thread-like through-
(1)   out most of their length (a), sometimes with broad segments in the middle (b), but never with a club at the end 3

Note: Distinguish between two types of head appendage:
  i) palps (mouthparts) which appear to have 3 segments;
  ii) antennae, with many more segments, usually 7–11. **Antennae and palps may be hidden beneath the head; see note on p. 5**

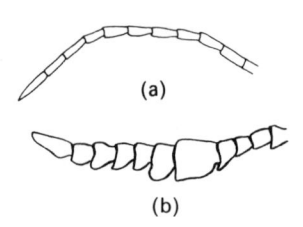

—   Antennae with a club at the end (c), or of a peculiar form (d), or apparently absent (see note above) 9

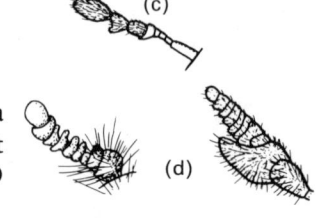

3	Antennae inserted on the front of the head, the distance between their bases less than the length of their first segment (a); all tarsi with 4 segments visible, the third segment deeply bilobed (b) or very small (c); some species with brilliant metallic coloration; more than 4 mm long . . . **CHRYSOMELIDAE (part)** Key E, p. 86	
(2)		

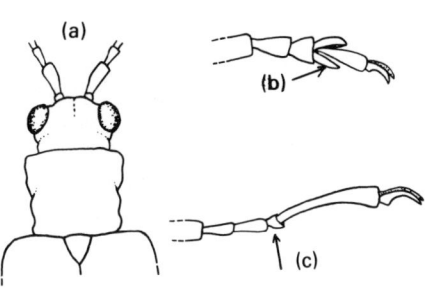

— Antennae inserted further apart than the length of the first segment; **hind** tarsi with 5 segments, either all simple, or with a lobe on the underside of segments 1–4, or with segment 4 bilobed (see next couplet); never with brilliant metallic colours, though a slight brassy reflection may be present; sizes 1.3–38.0 mm 4

4 Hind tarsal segments 1–3 without lobes,
(3) fourth segment bilobed (a); elytra soft and
   hairy . . **SCIRTIDAE (Helodidae)**
                              (See Joy, 1932)

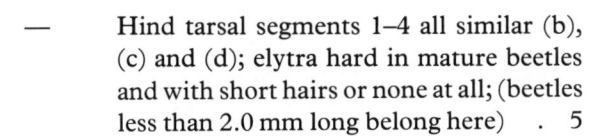

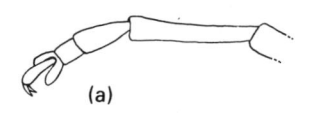

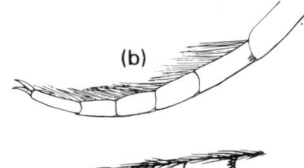

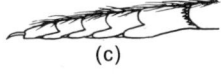

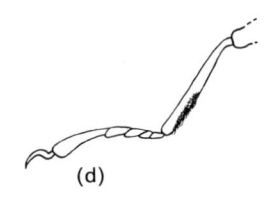

— Hind tarsal segments 1–4 all similar (b), (c) and (d); elytra hard in mature beetles and with short hairs or none at all; (beetles less than 2.0 mm long belong here) . 5

5 (4)	Pronotum as long as wide; head and eyes almost hidden below the convex front margin of the pronotum (a) (beware specimens preserved in alcohol where the head may be extended); hind tarsi with the last segment very long and bulbous at tip (b); size range 1.3–4.75 mm . . . . . . **ELMIDAE (Elminthidae)** Key F, p. 88
—	Pronotum wider than long; head and eyes clearly visible from above (c); hind tarsi tapering towards the tip, with or without a fringe of long pale swimming hairs; size range 1.7–38.0 mm 6
6 (5)	Elytra each with about 10 longitudinal rows of large pits (punctures) (a); hind coxae with large, rounded plates which cover half of the abdomen and the basal half of the hind femora (b); size range 2.5–5.0 mm . . . . **HALIPLIDAE** Key B, p. 22
—	Elytra without 10 rows of punctures, but 1 to 5 longitudinal lines or grooves may be present; hind coxae may have lobed or pointed projections, but never large rounded plates; most of the hind femora visible (c); size range 1.7–38.0 mm . . 7

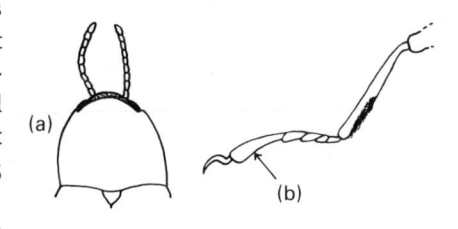

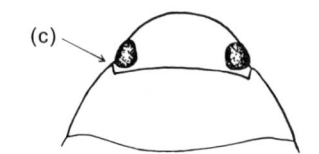

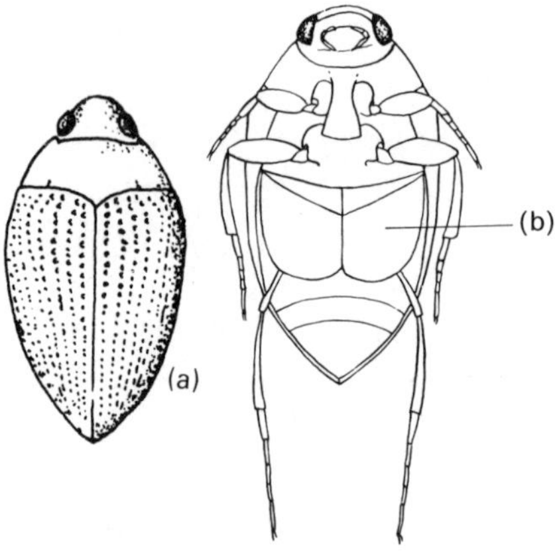

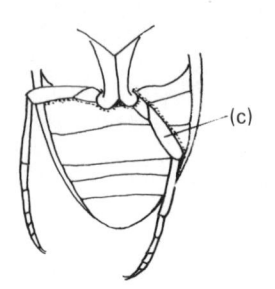

7	Head narrower than the front of the pronotum; beetle 8.5–10 mm long, strongly convex below; elytra black with yellow or red front and side margins (a); may squeak when alarmed: "screech beetle". . . . **HYGROBIIDAE** 1 sp. **Hygrobia hermanni**
(6)	

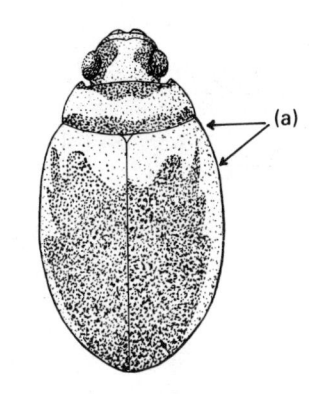

— Head about the same width as the front of the pronotum and usually inserted into it, so that the front margin of the pronotum and the head is a smooth curve (b) (beware specimens preserved in alcohol where the head may be extended beyond its normal position); beetle of different combination of size, convexity and coloration . . 8

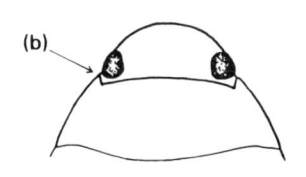

8	Hind coxal processes very broad (together wider than long) with a W-shaped hind margin (a), and overhanging the rest of the body at the sides (b); elytra widest near front, glossy brown with scattered large punctures concentrated towards the rear (c); middle antennal segments cup-shaped, with lobes in the male (d); size 3.5–5.0 mm . . . . **NOTERIDAE** Key C, p. 30
(7)	

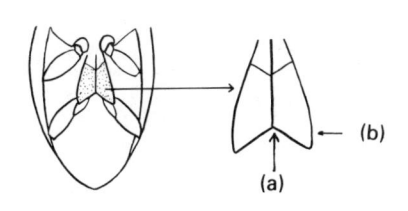

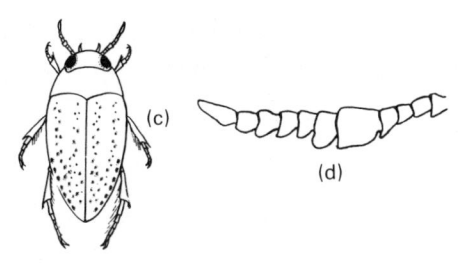

— Hind coxal process longer than broad, with a lobed (e), pointed, straight or wavy hind margin (f); in most species, the elytra are widest further back and punctures are not concentrated at the rear; middle antennal segments usually longer than wide and never with lobes; size 1.7–38.0 mm . . . . . **DYTISCIDAE** Key D, p. 31

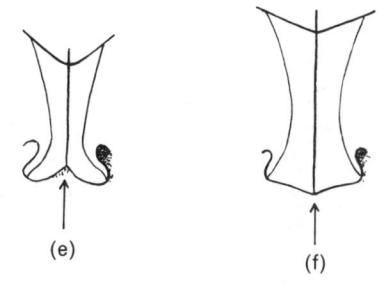

9 (2)	Head with an elongate, drooping snout (a), at least as long as broad (look from the side as well as from above); antennae long, with a definite "elbow" between the first and second segment (b) in most species . **CURCULIONIDAE** (weevils) (See Joy, 1932 for keys to species)	
—	No such snout, though the front of the head may bulge out in the centre; antennae without an "elbow". 10	
10 (9)	Front and middle legs longer than the entire beetle; antennae short **ELMIDAE** Key F, p. 88	
—	Front and middle legs shorter than the entire beetle; antennae long or short . 11	
11 (10)	Elytra and pronotum covered with dense hair; sizes 2.5–5.9 mm 12	
—	Elytra and pronotum without hairs, or with sparse hairs; sizes 1.0–48.0 mm . 13	
12 (11)	Last segment of the tarsi long; all tibiae slender; antennae as shown (a); elytra of uniform colour . . . **DRYOPIDAE** Key H, p. 141	
—	All tarsal segments short; front tibiae very broad; antennae as shown (b); in mud around ponds and saltmarshes; elytra usually with a distinctive pattern (c) . . **HETEROCERIDAE** [1 genus *Heterocerus,* 8 species] (See Clarke, 1973 for a key to species)	

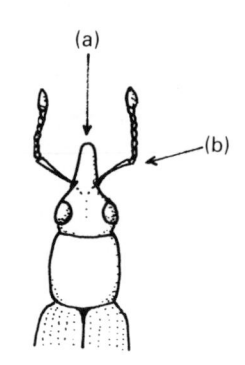

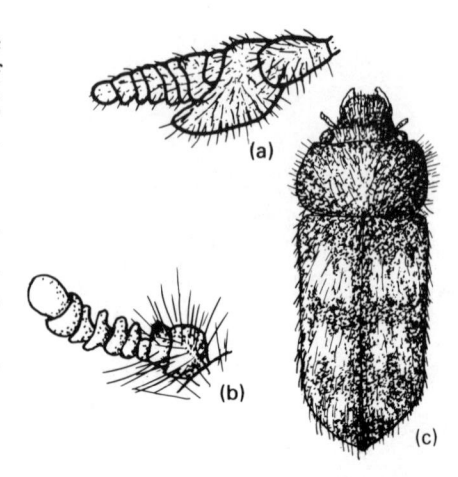

13 (11)	Palps from $\frac{3}{4}$ to 4 times as long as the antennae and usually obvious; antennal club with 3–5 enlarged, hairy segments (a); sizes 1.0–48.0 mm . . . . **HYDROPHILIDAE & HYDRAENIDAE** Key G, p. 93	(a)

Note: Distinguish between two types of head appendage:
  i) palps (mouthparts) which appear to have 3 segments;
  ii) antennae, with many more segments, usually 7–11. **Antennae and palps may be hidden beneath the head; see note on p. 5**

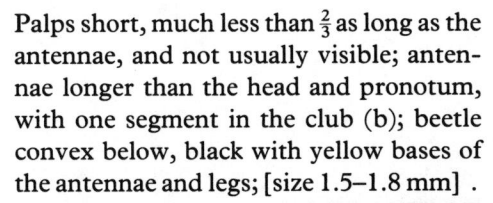

— Palps short, much less than $\frac{2}{3}$ as long as the antennae, and not usually visible; antennae longer than the head and pronotum, with one segment in the club (b); beetle convex below, black with yellow bases of the antennae and legs; [size 1.5–1.8 mm] .
LIMNICHIDAE
[1 sp. ***Limnichus pygmaeus***]

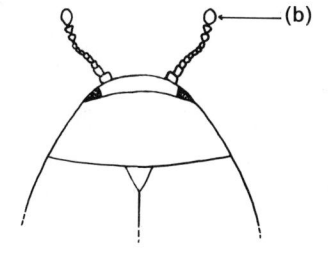

KEY A GYRINIDAE

Family Characters: hind and middle legs short and broad; eyes divided.
  "Whirligig" beetles.
  Size range: 3.5–7.8 mm.
  2 genera, 12 spp.

Some species are difficult to separate with certainty except by comparison of the genitalia; where the protruding tips provide useful characters, these are drawn; for complete figures, see Foster, 1981. Males can be recognised by the sucker hairs on the front tarsi, and by the three parts of the protruding genitalia.

1   Pronotum and elytra covered with fine, short hairs; labrum bulging, not forming a continuous curve with the front of the head (a) [size 5.5–6.5 mm]. . . . . .
  Orectochilus villosus

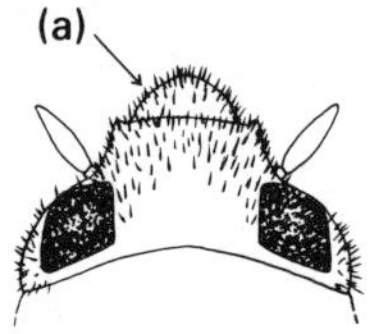

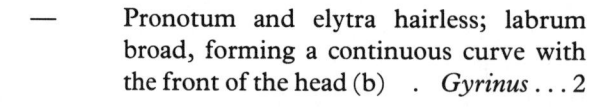

—   Pronotum and elytra hairless; labrum broad, forming a continuous curve with the front of the head (b) . *Gyrinus* . . . 2

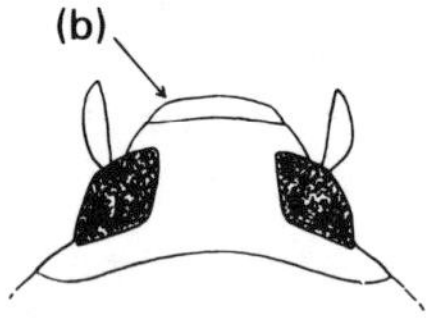

2   Underside entirely yellow-red . . . 3
(1)

—   Underside mainly dark, with some parts orange 4

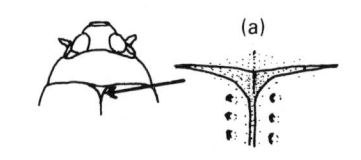

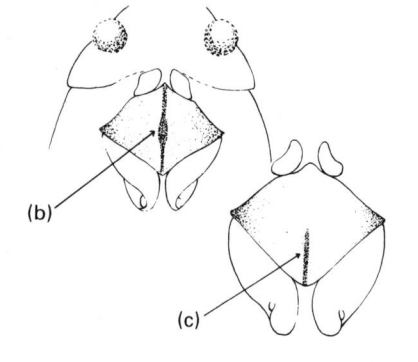

3   Size 3.5–4.5 mm; mesoscutellum (at junction of pronotum with elytra) with a short longitudinal ridge (a); mesosternum (between middle and front coxae) with a central groove running full-length (b); elytra metallic blue, not striped . . . .
(2)   **G. minutus**

—   Size 5.0–7.8 mm; no ridge on mesoscutellum; mesosternal groove confined to the hind half (c); elytra with stripes of metallic green and blue . . . . **G. urinator**

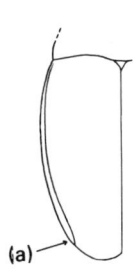

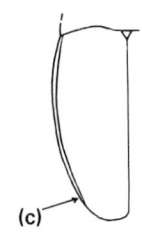

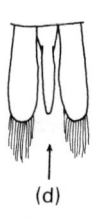

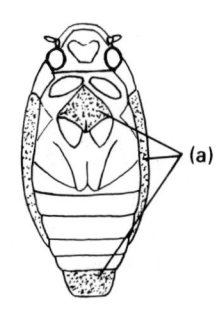

4 (2)	Middle and hind tarsal claws black, much darker than the yellow legs and tarsi . 5	
—	Middle and hind tarsal claws yellow, same as the legs 6	
5 (4)	Flattened lateral rim of the elytra broadening out behind the widest point of the elytra (a) (view from above), male genitalia diagnostic: penis needle-sharp (b) [size 4.5–7.5 mm] . . . **G. marinus**	
—	Rim of the elytra of a constant width, or narrowing slightly behind the widest point of the elytra (c); penis blunt at tip (d) [size 4.5–6.3 mm] . . . . **G. aeratus**	
6 (4)	Pronotum and elytra dull all over and clearly reticulated (at ×20); underside generally dark all over except for the epipleurs (see Fig. 3) [size 5.0–6.5 mm] . **G. opacus**	
—	Pronotum and elytra shining, with or without weak reticulation; underside usually with the metasternum and/or the last abdominal segment orange in addition to the epipleurs (a) 7	

| 7 (6) | Elongate beetles with sides of the elytra almost parallel in the middle; elytra scarcely wider at their mid-point than at their front margin (a)(b) . . . . . 8 |

| — | Rounded beetles; elytra at widest point about 1.25 times wider than at the front margin (c) . . . . 9 |

| 8 (7) | Male genitalia with tip of the penis broad, about 2/3 the width of an outer lamella (a); female genitalia with ends of the ovipositor lobes sloping outwards (b) [size 5.0–7.5 mm] . . . .***G. caspius*** |

| — | Penis narrow, about 1/5 the width of a lamella (c); end of the ovipositor lobes sloping inwards (d) [size 5.5–7.8 mm] . . ***G. bicolor*** |

| 9 (7) | Lens-shaped group of punctures at the tip of each elytron deeply impressed, larger and stronger than the rest of punctures (a) (tilt the dry beetle to and fro in the light with the hind end higher than the head); penis narrow or broad, considerably shorter than the outer lamellae (males) (b) . . . . 10 |

| — | Lens-shaped group of punctures on the elytra faint (fainter than other punctures) (c); penis narrow, almost as long as the lamellae (males) (d) [size 4.0–6.2 mm] . . ***G. suffriani*** |

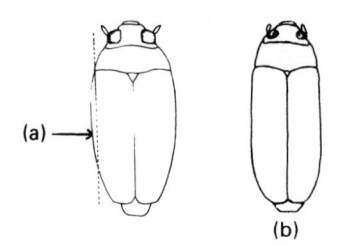

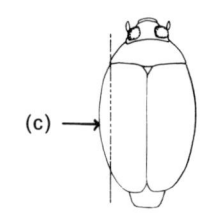

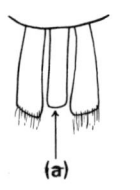

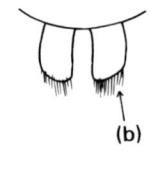

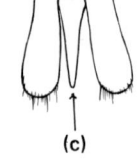

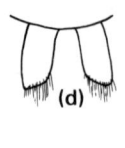

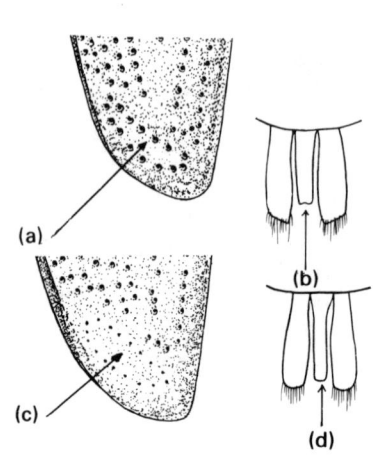

10	Male genitalia with tip of the penis almost as wide as an outer lamella (a); female genitalia with outer edge of the ovipositor lobes strongly curved (b) [size 5.0–7.0 mm] . . . . . . *G. distinctus*
(9)	

—	Penis narrowing to about 1/2 width of a lamella at its tip (c); ovipositor lobes cut off more squarely at the tip (d) [size c.6.0 mm] 11

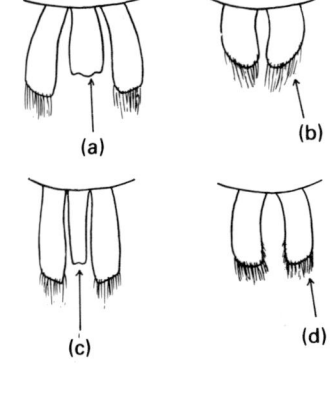

11	Ovipositor lobes of female (extract genitalia from abdomen—see p. 5) with distinct tooth at inner corner (a) . . *G. natator*
(10)	

—	Ovipositor lobes without a distinct tooth (b) . . . . *G. substriatus*

Note: *G. natator* tends to be black on the last abdominal segment and *G. substriatus* orange. True *G. natator* occurs in Ireland but is known to have occurred in Britain on only a few occasions (see Angus & Carr, 1982).

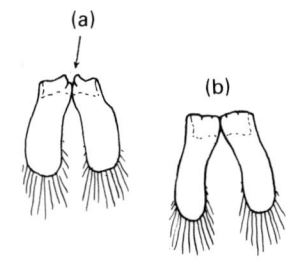

KEY B HALIPLIDAE

Family Characters: Long hind legs, with tapering tarsi of 5 segments bearing swimming hairs; hind coxal plates cover bases of femora; elytra with longitudinal rows of large punctures.
Size range 2.5–5.0 mm.
3 genera.

1   Strong longitudinal ridges on the elytra (a); pronotum bulging out behind the anterior margin . . ***Brychius elevatus***

—   No ridges on the elytra; pronotum tapering towards the anterior margin . . . 2

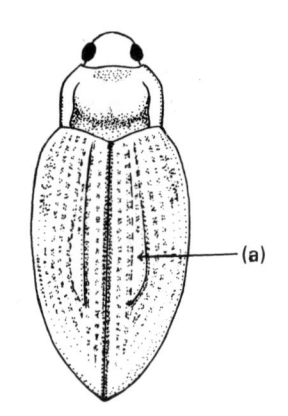

2   Hind coxal plates each with a point on the hind margin (a); last segment of maxillary
(1)   palp (the longer pair of palps) longer than the preceding one (b) ***Peltodytes caesus***

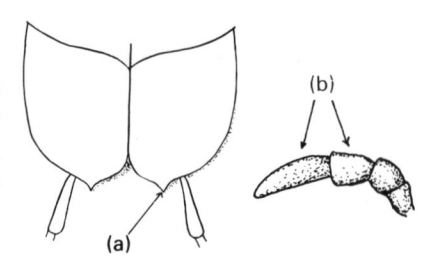

—   Hind coxal plates with a smoothly rounded hind margin; last segment of maxillary palp shorter than the preceding one (c) . . . . ***Haliplus***
                                   Key B1

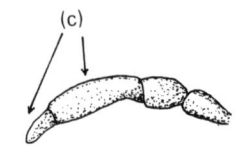

Key B1   ***Haliplus***

Genus Characters: Hind coxal plates with smooth, rounded hind margin.
                 17 spp.
                 Size range 2.5–5.0 mm

Note: a number of species of *Haliplus* can be identified with certainty only by reference to the male genitalia. Figure 7 (p. 29), at the end of the key gives the form of the aedeagus (middle part of the genitalia) for the males of all the species keyed out here, and the left (more pointed) paramere for the *ruficollis* group. The figures are redrawn from

Balfour-Browne (1940), Parry (1982) and from fresh material. Male *Haliplus* can be distinguished from females by the wide basal segments of the front tarsi and the pale sucker hairs beneath the segments (see couplet 10). See note p. 5 on extracting the genitalia.

1   Pronotum with a long, curved furrow on each side, marked by a distinct dark line and extending almost half way to the front margin (a); puncture rows at the front edge of the elytra developed into raised ridges (b); the side rim of the pronotum wide at the rear end, about as wide as the width of an antennal segment (c); most specimens have a black line down the centre of the pronotum (d); aedeagus (1) [size 2.6–3.5 mm] . .**H. lineatocollis**

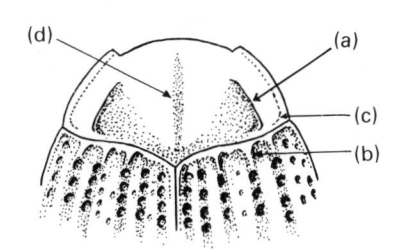

—   Pronotum without furrows or with short furrows extending no more than 1/3 the way across (e); no ridges on the front of the elytra (though the punctures may be enlarged here); the side rim of the pronotum always narrower than an antennal segment; no black line down the centre of the pronotum. 2

     Note: Furrows, if present, occur adjacent to the 4th or 5th row of punctures of the elytra and consist of a shallow elongate depression marked on the outer edge by a sharply-defined edge or slight ridge. View with the light from the side on a DRY beetle.

     Beware: Dark shadow marks which may be present on the hind margin of the pronotum above the 3rd rows of punctures from the centre.

2   Pronotum with a broad black band along
(1)   the hind margin and another along the front margin (a); elytral pattern as shown (b); aedeagus (2); [size 2.6–2.9 mm] . . **H. varius**

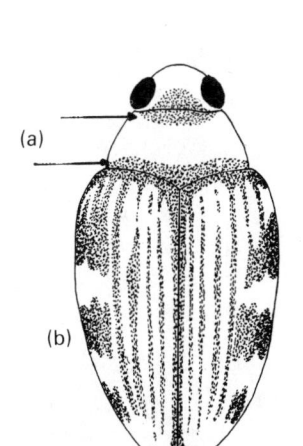

—   Pronotum without black bands, although some diffuse dark pigment and the shadow of the head may be visible along the front edge. 3

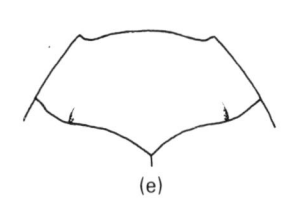

3 (2)	Pronotum, elytra and entire underside covered with very small pits which are visible on a dry beetle at × 20 as individual pin-prick dots—these small punctures are in addition to those forming rows on the elytra and those about a femur's width apart on the hind coxal plates (a); no large punctures on the epipleurs (the sides of the elytra seen from below) (b) . . . 4	
—	No fine punctures visible on the pronotum or on the hind coxal plates and none obvious on the elytra except at magnifications above × 40 on some females; epipleurs with large punctures, the same size as those on the hind coxal plates (c) 5	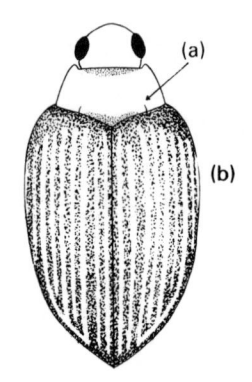
4 (3)	Hind margin of the pronotum with short furrows (a); elytra usually with continuous darker lines along puncture rows (b); aedeagus (3); [size 3.0–3.5 mm] . . . . **H. confinis**	
—	No furrows on the hind margin of the pronotum; elytra with the dark lines interrupted to produce blotches (c); aedeagus (4); [size 3.0–3.5 mm] . . **H. obliquus**	

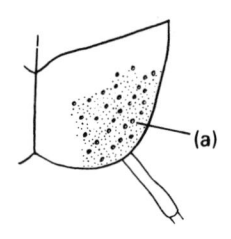

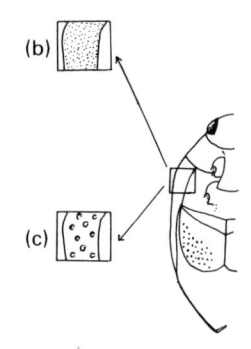

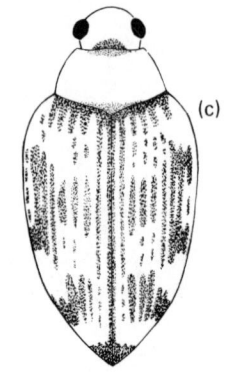

5 (3)	No furrows on the hind margin of the pronotum; size 2.5–5.0 mm 6	
—	Short furrows, extending between 1/10 and 1/3 the way across the pronotum, visible at ×20 (see note at cpt 1); size 2.5–3.2 mm 10	
6 (5)	Side borders of the prosternal process (marked by a sharp edge) not extending to the front of the prosternum (a) (tip the head up slightly and illuminate from the side); aedeagus (5); [size 3.5–4.0 mm] . . **H. flavicollis**	
—	Side borders of the prosternal process extending to the front of the prosternum (b) 7	
7 (6)	Hind edge of the pronotum narrower than the front edge of the elytra, so that the pronotum and elytra meet with a step, the angle between them about a right angle, seen from directly above (a); male (see note on sexing at beginning of key) with the basal segment of the middle tarsus scooped out at the tip (b); aedeagus (6); [size 2.5–3.0 mm] . . **H. laminatus**	
—	Hind edge of pronotum not narrower than the front edge of the elytra, the two parts meeting at a wide angle (c); male with the basal segment of the middle tarsus not scooped out at the tip 8	

| 8 | Elytra with dark flecks or blotches (see next couplet); the raised area just behind the middle coxae with a shallow pit at the point where the level falls away towards the hind coxae (a) (light from the side, turn the dry beetle in the light, view at ×20 or more); [size 2.5–4.2 mm] . . . . . 9 |
|(7)| |

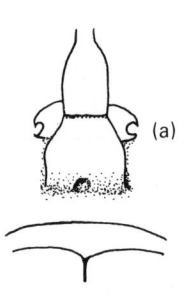

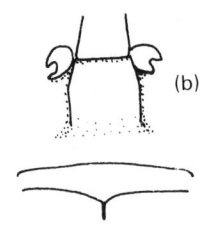

— Elytra uniformly dark yellow brown, except for the punctures, which are darker; no pit in the position described, the end of the prosternal process and the area behind the middle coxae all flat, in the same plane (b); aedeagus (7); [size 4.0–5.0 mm] . . . . **H. mucronatus**

9 Elytra usually with small dark markings,
(8) each covering 1 or 2 rows of punctures (a); sides of the elytra bulging at the front; aedeagus (8); [size 3.5–4.2 mm] . . . . **H. fulvus**

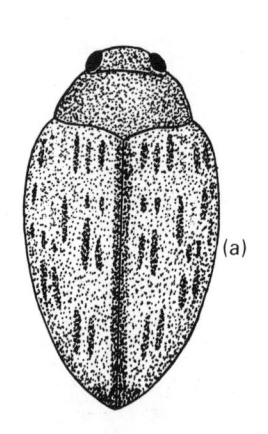

— Elytra usually with extensive dark markings, each covering 3–4 rows of punctures (b); sides of the elytra and the pronotum with a continuous outline; aedeagus (9); [size 2.5–3.5 mm] . . **H. variegatus**

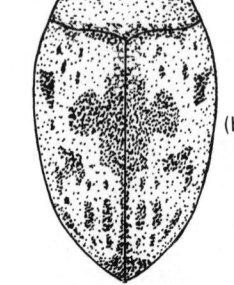

Note: H. fluviatilis could come out here if the very short furrows on the pronotum have been missed; see table at end of key for details of the male genitalia and the characteristic elytral pattern.

10 (5)	**The *ruficollis*-group: rely on identification of males and refer to the table at the end of the key for details of the male genitalia.**	
—	Front tarsi with the basal segments enlarged below, and bearing tufts of pale hairs on the underside (a); basal segment of the middle tarsi swollen, compared with the 2nd segment . . . . 11 **Males—*ruficollis* group**	(a)
—	Front tarsi with all segments symmetrical, without tufts of pale hair beneath (b); basal segment of middle tarsi not swollen **Females—*ruficollis* group**	(b)
11 (10)	Front tarsus with one claw about 2/3 the length of the other and more strongly curved (a) 12	(a)
—	Front tarsus with the two claws about the same length and curvature (ratio of claw lengths from 1:1 to 4:5) (b) 13	(b)
12 (11)	Underside of the basal segment of the **middle** tarsi very clearly concave (a) (depth of concavity much greater than the diameter of tibial spur); aedeagus (10); paramere type d . . ***H. immaculatus***	(a)
—	Underside of the basal segment of the middle tarsi almost straight (b) (depth of any concavity less than the diameter of tibial spur). . . . . males of ***H. ruficollis*** aedeagus (11) paramere type a ***H. wehnckei*** aedeagus (12) paramere type b	(b)

13 (11)	Underside of the basal segment of the **middle** tarsi very clearly concave (see cpt 12); aedeagus (13); paramere type a. . . **H. lineolatus**

—	Underside of the basal segment of the middle tarsi almost straight . . . . .
	  males of **H. heydeni** aedeagus (14)
	       paramere type c
	    H. apicalis aedeagus (15)
	       paramere type e
	  H. fluviatilis aedeagus (16)
	       paramere type a
	  H. furcatus aedeagus (17)
	       paramere type e

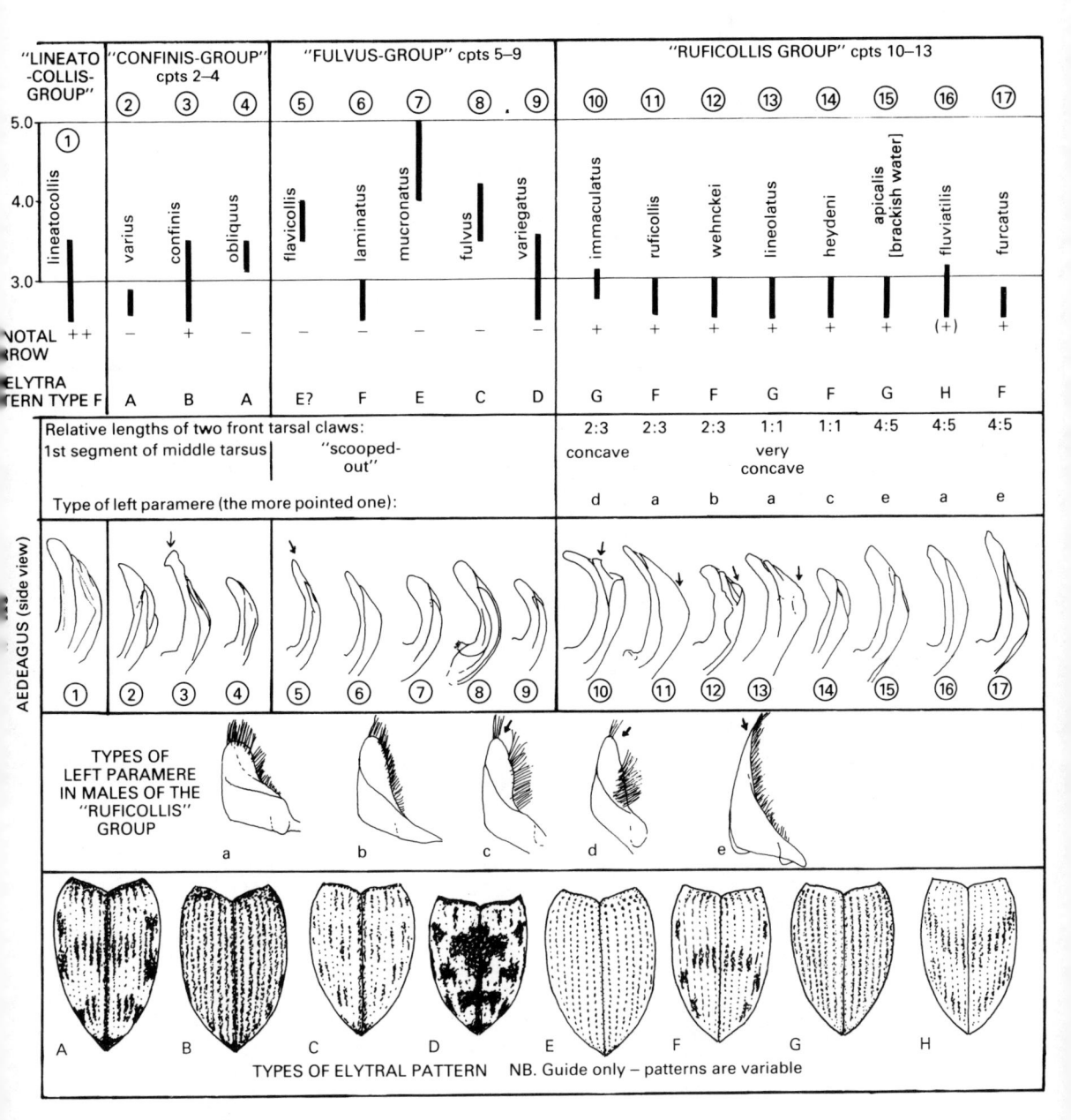

Figure 7. Tabular guide to *Haliplus* spp. To be used in conjunction with Key B1, p. 22.

KEY C NOTERIDAE

Family Characters: hind coxal process wider than long, overhanging coxae at sides; middle antennal segments cup-shaped, lobed in male.
1 genus Noterus *2 spp.*

1   Prosternum with a ridge in the mid-line, running from a minute point on the front margin back along the process; end of the process with a narrow "neck" (a); [size 4.0–5.0 mm] . . . . ***N. clavicornis***

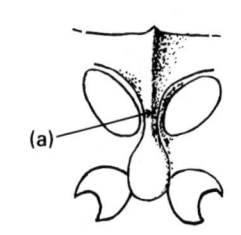

—   Prosternum without a mid-line ridge; front margin of the prosternum smooth; process with a wide "neck" (b); [size 3.5–4.0 mm] . . . . . ***N. crassicornis***

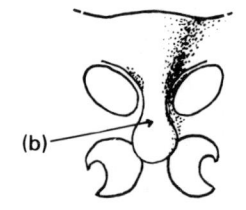

Note: The larger sp. was previously known as *N. capricornis* and the smaller as *N. clavicornis*; to avoid confusion, it is worth adding "the larger species" after naming *N. clavicornis*, and "the smaller species" after *N. crassicornis*.

KEY D   DYTISCIDAE

Family Characters: Long hind legs with tapering tarsi of 5 segments, bearing swimming hairs; hind coxae with rounded, pointed or truncate processes.
Size range 1.5–38.0 mm.
4 sub-families.

Note: **This key is based on size and colour patterns. The box at the end should be checked to confirm the correct identification of the sub-families Dytiscinae and Colymbetinae.**
See also Tables 1 and 2, pages 142–145, for a guide based on the coloration of species of Colymbetinae/Dytiscinae and Hydroporinae respectively.

1	Small triangular plate (mesoscutellum) visible at the junction of the pronotum with the elytra (a); size 6.0 mm or more	2
—	No mesoscutellum visible (b), but pronotum may be extended backwards into a point in the mid-line; size 6.0 mm or less	7
2 (1)	Size 22 mm or more. . . **Dytiscinae**	Key D2, p. 50
—	Size 18 mm or less	3
3 (2)	Main area of the elytra mottled (a) with a strong yellow and black pattern (use ×6 or more)	4
—	Main area of elytra a uniform colour or only faintly mottled, with or without pale stripes or bars	5

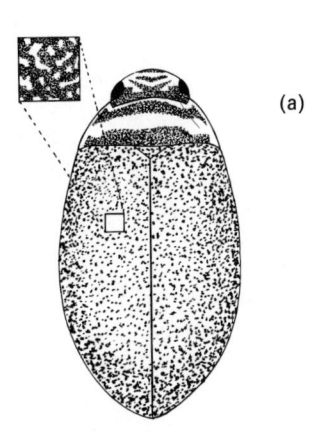

4 (3)	Size 14 mm or more. . .	**Dytiscinae** Key D2, p. 50
—	Size 12 mm or less . .	**Colymbetinae** Key D1, p. 33
5 (3)	Pronotum pale with a dark crescent mark on the hind margin (a) 6	(a)
—	Pronotum with other markings, or of a uniform colour . . .	**Colymbetinae** Key D1, p. 33
6 (5)	Size 12 mm or more. . .	**Dytiscinae** Key D2, p. 50
—	Size 9 mm or less . .	**Colymbetinae** Key D1, p. 33
7 (1)	**Hind** tarsi with broad segments with long lobes underneath (a); **front** and **mid** tarsi with 5 simple segments **Laccophilinae** Key D3, p. 55	(a)
—	Segments of the hind tarsi narrow and simple; front and mid tarsi with apparently 4 segments, the 3rd deeply lobed . **Hydroporinae** Key D4, p. 56	

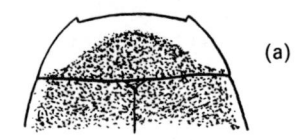

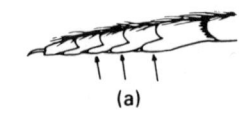

Confirmatory characters for distinguishing Dytiscinae and Colymbetinae

Eyes, seen from in front, with a strong indentation just above the antennae (a); males with pale sucker hairs on the broad segments of the front tarsi (b); females never with deep fluting on the elytra. Size range, 6.0–18.0 mm. . **Colymbetinae**
Key D1, p. 33

Note: Tip the beetle up through 90° so that you can look at the very front of its head, face on. (You may need to steady its back end on a small blob of plasticine or Blu-tak).

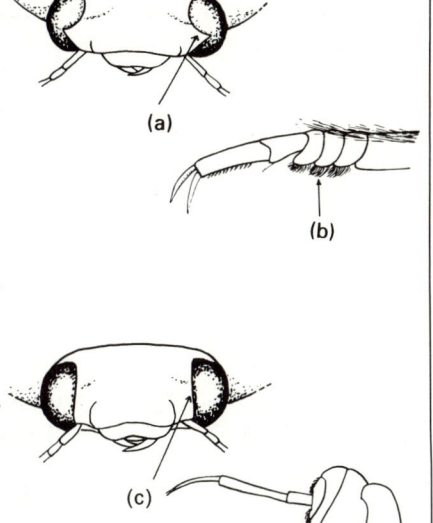

Eyes with no indentation or only a very slight one (c); males with segments 1–3 of the front tarsi broadened to form a circular plate bearing cup-shaped pale suckers beneath (d); females may have deep fluting on the elytra. Size range, 12.0–38.0 mm . . . . . . **Dytiscinae**
Key D2, p. 50

Key D1 **Colymbetinae**

Sub-family Characters: mesoscutellum visible; indentation on front of eyes.
Size range 6.0–18.0 mm.
6 genera.

Note: Table 1 (p. 142) is an additional guide based on the colours of Colymbetinae and Dytiscinae species.

1   Size 16.0–18.0 mm; elytra with a distinct transverse reticulation (a) (a drop of water will run sideways off the beetle's back) .
Colymbetes fuscus
(See front cover illustration)

—   Size usually 12.0 mm or less, or, if larger, colour uniform black; reticulation on the elytra with a square or longitudinal mesh (a drop of water will run backwards off the beetle's back or stay put) 2

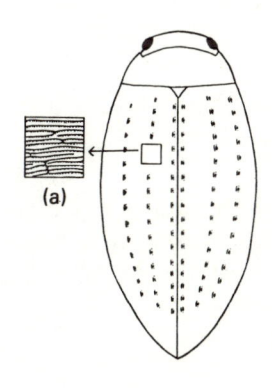

2 (1)	Epipleurs of the elytra (underside) remain broad almost to the tip of the abdomen (a); elytra dark brown, usually with yellow markings (b); [size 7.5–8.5 mm] . . . . *Platambus maculatus*	
—	Epipleurs get much narrower at the level of first abdominal segment (c) . . . 3	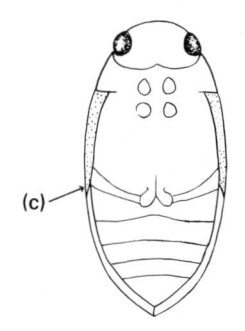
3 (2)	Lobes of the hind coxal process diverge widely; lines on the hind coxal process virtually touching in the mid-line then turning sharply through almost 90° to run onto lobes (a); body elongate, elytra uniform red; [size 7.0–8.0 mm] . . . . *Copelatus haemorrhoidalis*	
—	Hind coxal lobes diverge only slightly; hind coxal lines always widely separate never turning so abruptly onto lobes (b) (c) 4	

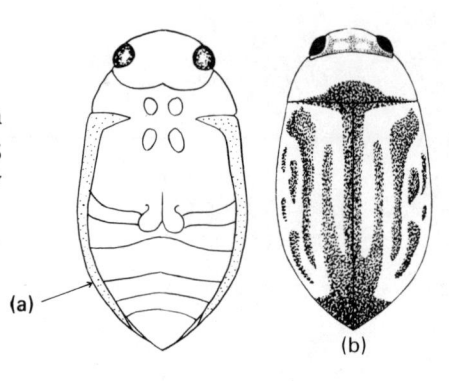

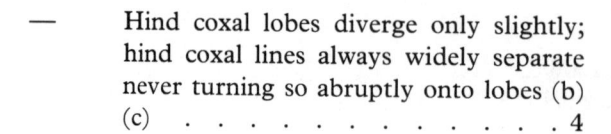

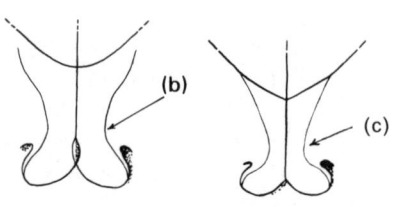

4	Hind femur without a comb (see below); pronotum yellowish with or without dark bands and the elytra mottled (a) (except in 1 sp. where the pronotum and elytra are entirely black); hind margins of the hind coxal lobes with a notch (b) . **_Rhantus_** Key D1a, p. 35
(3)	

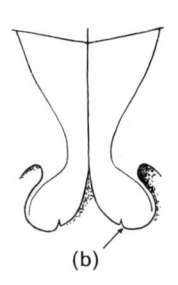

—	Hind femur (c), seen from below, with a comb of 3–8 spines on the lower hind corner (these spines may be strong and neatly lined up (d), or weak and splayed (e)); pronotum and elytra usually with different colour combinations to above; hind coxal lobes with or without a notch. . . . **_Ilybius_ and _Agabus_** Key D1b, p. 38

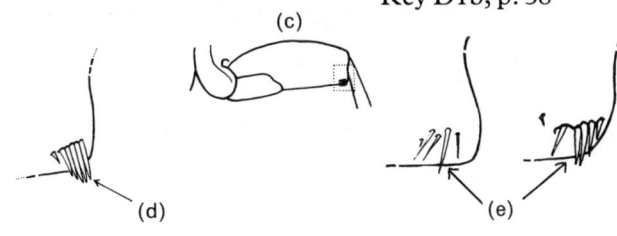

Key D1a   *Rhantus*

Genus Characters: no hind femur comb; hind coxal process with notch.
            Size range 9.0–12.0 mm.
            6 spp.

Note: males are distinguished from females by the presence of pale sucker hairs on the underside of the front tarsi.

1	Elytra and the pronotum black; [size 10.0–11.0 mm] . . . . . **_R. grapii_**
—	Elytra mottled yellow and black (× 6) . 2

2	Pronotum with a black mark in the centre; with (a) or without (b) dark mark on hind edge of the pronotum; males with the two front tarsal claws very unequal in length, one no more than 2/3 as long as the other (see cpt 3); [size 10.0–12.0 mm] . . . 3
(1)	

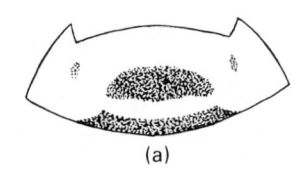

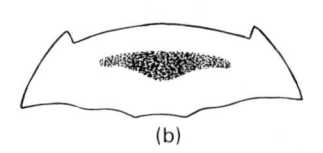

—	Dark marks on pronotum, if any, confined to the hind edge (c) (d); males with the front tarsal claws only slightly unequal (see cpts 4–5); [size 9.0–10.0 mm] . . 4

3	Black band on hind edge of pronotum; central mark usually split into three (a); 2–3 yellow lines running down each elytron; male abdomen (underside) with dark segments which are paler at the sides, female abdomen vice versa; male with the longer front tarsal claw longer than the last tarsal segment (b); [size 10.0–11.0 mm] . **R. frontalis**
(2)	

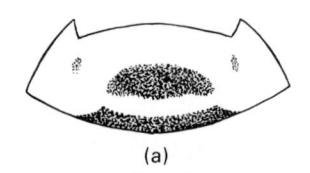

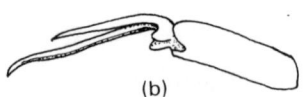

—	No black band on hind margin of pronotum (c); no yellow lines down elytra; abdomen of both sexes all black (except in soft immature specimens); male with longer front tarsal claw shorter than last tarsal segment (d); [size 10.0–12.0 mm] . **R. suturalis**

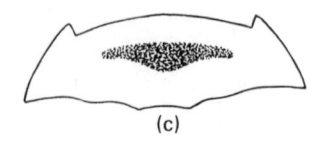

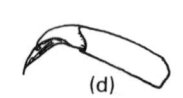

4 (2)	Abdomen uniformly yellow-brown (or, rarely, dark brown) on underside; dark markings on hind edge of the pronotum rather diffuse, not constituting a black band (a); male with very long front tarsal claws, both much longer than the last tarsal segment, and serrated (b); [size 9.0–10.0 mm]. . . . . . **R. exsoletus**	
—	Abdomen predominantly black on underside; distinct black band present on hind edge of pronotum (c); male front tarsal claws as long as, or only a little longer than, the last tarsal segment (see cpt 5) 5	
5 (4)	Abdomen all black; 2–3 black lines down each elytron; male front tarsal claws both longer than the last tarsal segment (a); [size 9.0–10.0 mm] . . **R. suturellus**	
—	Abdominal segments black with yellow sides; no black lines down elytra; shorter male front tarsal claw a little shorter than the last tarsal segment (b); [size c.9.5 mm] **R. aberratus**	

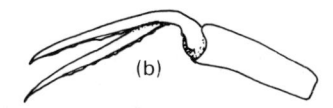

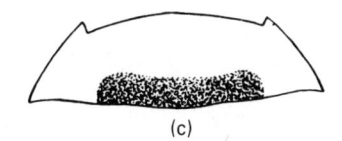

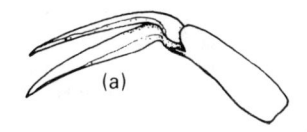

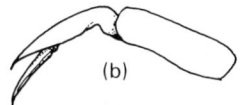

(Figures of male claws redrawn from Balfour-Browne, 1950).

Key D1b   ***Agabus* and *Ilybius***

Agabus: Size range 6.0–11.0 mm.
            20 spp.

Ilybius: Size range 9.0–15.0 mm.
            7 spp.

READ THESE NOTES BEFORE PROCEEDING:

i The prosternal process.
   The first segment of the underside of the thorax is extended backwards in the midline into a process (prosternal process) (a) which reaches to the middle coxae. The process is spear-shaped, consisting of a "neck", which has a longitudinal ridge, and a widened end part (b). This **end part** may have a **ridge or arch in cross-section, or may be flattened at its broadest point,** in which case a definite change in cross-section is apparent at the junction with the "neck". View the beetle **dry, illuminating from the side.**

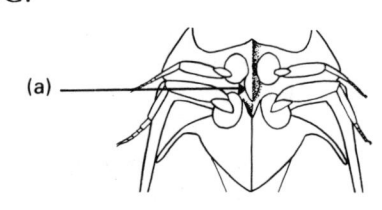

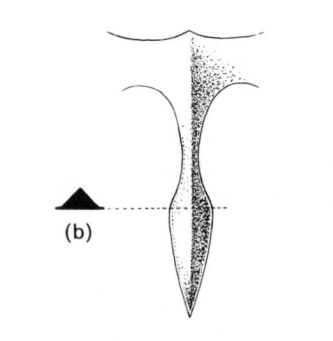

ii Male characters.
   Check for presence of sucker hairs on the underside of segments 1–3 in the front and mid tarsi (c) to confirm that the beetle is male.

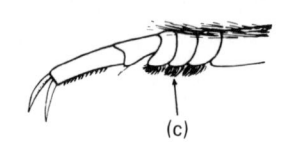

Figure 8, p. 49, shows the sizes, colour patterns and prosternal process characters of all the species of *Agabus* and *Ilybius* keyed out here.

1   Size 13.0–15.0 mm; beetle all black. . . .
                                    Ilybius ater

—   Size 6.0–11.5 mm; beetle black, brown or brassy, with or without patterned elytra and pronotum 2

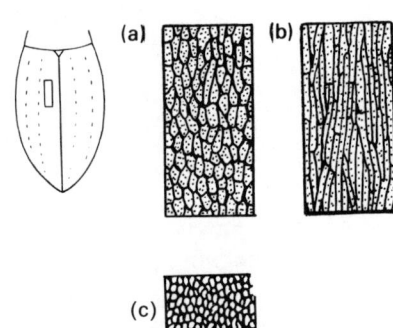

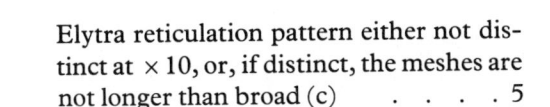

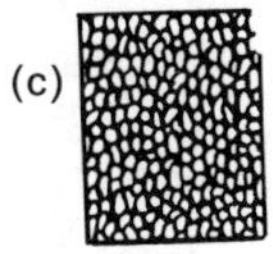

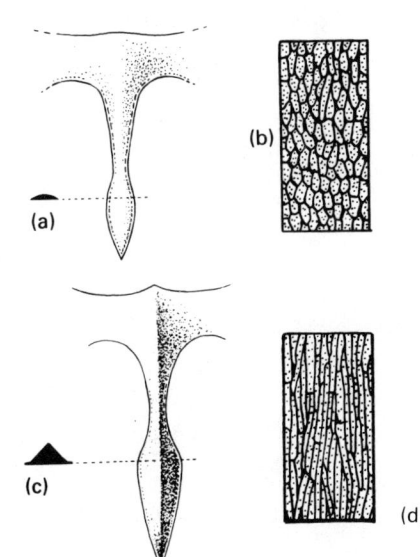

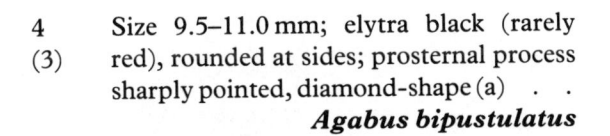

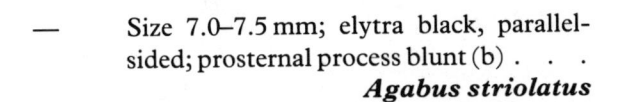

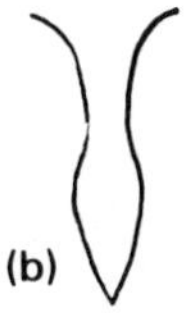

2 (1)	Elytra with reticulation (a net-like pattern), distinct at × 10 and with the meshes **longer than broad** ((a) and (b) show minimum and maximum elongation of meshes)—light from the side reveals a fine "cracked" appearance on a **dry** beetle.	3
—	Elytra reticulation pattern either not distinct at × 10, or, if distinct, the meshes are not longer than broad (c)	5
3 (2)	End part of prosternal process at most moderately arched (a), elytra black with a red streak on the lateral margin near the tip; reticulation meshes are only slightly longer than broad (b) [size 8.0–8.5 mm] . **Agabus melanarius**	
—	Prosternal process with a ridge along entire length (c); reticulation meshes are much longer than broad (d)	4
4 (3)	Size 9.5–11.0 mm; elytra black (rarely red), rounded at sides; prosternal process sharply pointed, diamond-shape (a) . . **Agabus bipustulatus**	
—	Size 7.0–7.5 mm; elytra black, parallel-sided; prosternal process blunt (b) . . . **Agabus striolatus**	

| 5 | Pronotum and elytra bronze with a broad, full-length, paler lateral stripe (a); underside yellowish; [size 10.0–11.0 mm] . . |
| (2) | ***Ilybius fuliginosus*** |

| — | No continuous broad paler stripe on pronotum and elytra (though the pronotum may have a pale border and the elytra may have streaks at the side margin in beetles less than 9.5 mm long); underside yellowish, reddish or black . . . . 6 |

| 6 | Elytra dark with conspicuous yellow pattern (see next couplet) . . . . . 7 |
| (5) | |

| — | Elytra uniform colour, or mottled all over, or dark with or without paler lateral streaks or spots . . . . 8 |

| 7 | Elytra with two pairs of lateral yellow spots (a); hind tibia stout—in broadest view only about twice as long as broad, and tibial spur (b) more than half the length of the tibia; head black; [size 7.0–8.0 mm] . . . ***Agabus didymus*** |
| (6) | |

Note: *A. biguttatus* and *A. guttatus* usually have reddish spots at the sides of the elytra; these two species are larger than *A. didymus* (c. 9.0 mm) and have a slender hind tibia–go to couplet 8.

| — | Elytra with yellow zig-zag pattern (c); hind tibia 3–4 times as long as broad, with the tibial spur less than half as long as the tibia (d); head reddish; [size 7.0–8.0 mm]. ***Agabus undulatus*** |

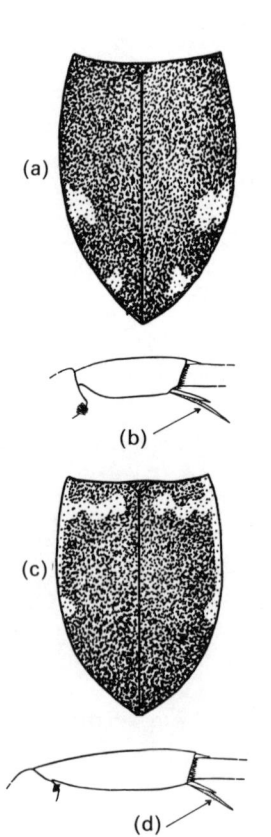

A Key to the Adults of British Water Beetles

8
(6)
Pronotum much paler than head, predominantly yellowish to mid-brown in contrast to predominantly black head; centre of elytra mottled 9

—
Pronotum predominantly dark, the dark areas as dark as the head, but may have paler side margins or pale bars across the centre; elytra uniform colour or mottled .
10

9
(8)
Elytra olive with black flecks; pronotum almost invariably with two black spots (a); femora usually uniformly pale; claws of the front tarsus of both males and females equal in length to the last segment of the tarsus (b); [size 8.0–8.5 mm] . . . . .
A. nebulosus

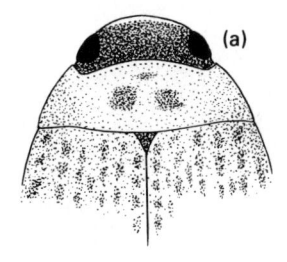

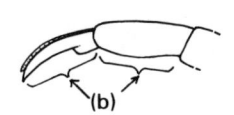

—
Elytra mottled brown; pronotum without two black spots; front and middle femora usually with a black patch (c); claws of the front tarsus of both sexes only about 2/3 length of the last segment of the tarsus (d) [size 8.0–8.5 mm] . . **A. conspersus**

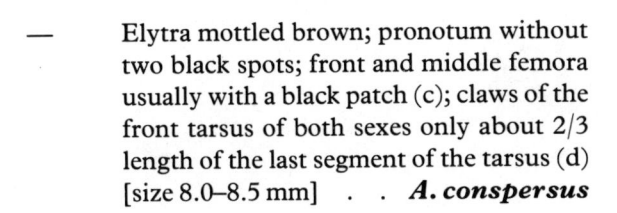

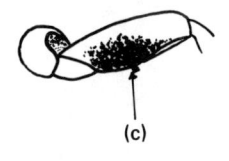

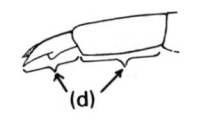

10 (8)	Pronotum with pale bars across (sometimes reduced to two spots) (a); end of the prosternal process narrowing abruptly and running out to a long point (b); end part of prosternal process flat at the **widest** point (c); [size 6.0–7.5 mm] . . ***Agabus arcticus***	
—	Pronotum uniform colour or dark with paler side margins; prosternal process not drawn out into a long point and flat, arched, or with a mid-line ridge at its widest point (d), (e), (f)—see note (i) at the top of the key (p. 38) 11	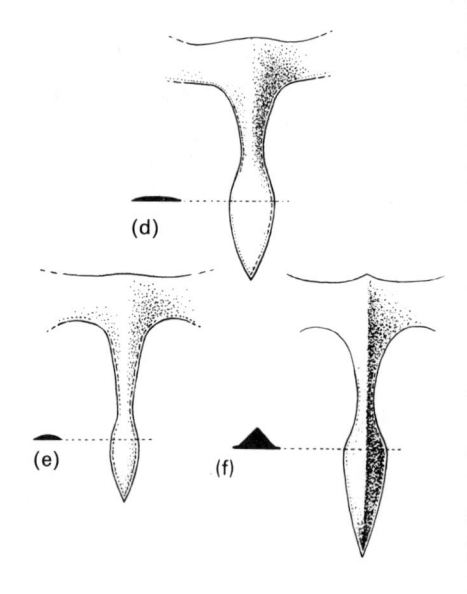
11 (10)	Elytra brown, paler than centre of pronotum; pronotum with conspicuous paler side borders (a) 12	
—	Elytra the same colour as the centre of the pronotum, brown, black or brassy; pronotum of uniform colour or with narrow paler side borders 14	

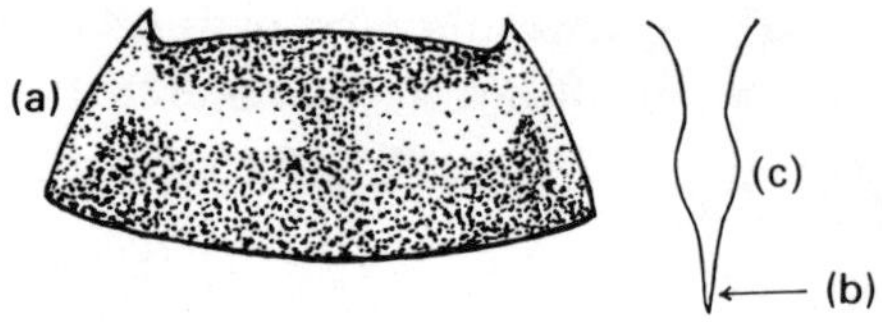

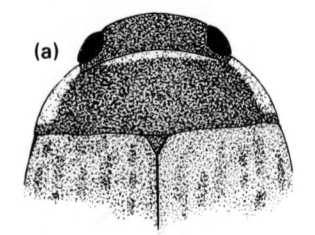

12	End part of the prosternal process flat in cross-section at its **widest** point (see couplet 10 fig. (d)); elytra olive flecked with black; both males and females (see note ii, p. 38) dull with net-like reticulation clearly visible on a dry beetle at × 10; [size 7.5–8.2 mm] ***Agabus sturmii***
(11)	

Note: If beetle has been kept in alcohol, allow to dry out before examining colours–the elytra may appear very dark unless dry.

— Prosternal process arched in cross-section at its widest point (see couplet 10 fig. (e)); elytra brown, uniform or mottled; males shiny with no reticulation obvious at × 10, and females shiny or dull, size c. 7.0 mm
. . . . 13

13	Elytra chestnut, pronotum dark brown; prosternal process broad and blunt (a); hind legs short (hind tibia when stretched out barely over-reaching the tip of the elytra); female as shiny as male; [size 7.0 mm] . . . ***Agabus paludosus***
(12)	

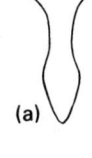

— Elytra dark brown, pronotum black; prosternal process narrow and pointed (b); hind legs long (tibia well over-reaching the tip of the elytra); female duller than male; [size 7.0 mm] . ***Agabus congener***

Note: Beware *A. labiatus* here; check the underside character as described in Couplet 21–*A. paludosus* and *A. congener* resemble Fig. (a), but *A. labiatus* resembles Fig. (c).

14 (11)	Elytra and the pronotum both chocolate brown; hind tibia stout (in broadest view, only twice as long as broad, with the longer tibial spur (a) more than half the length of the tibia); prosternal process flat at its **widest** point (b)—see note (i), p. 38 [size 9.0–9.5 mm] . ***Agabus brunneus***	
—	Combination of size and coloration different from above; hind tibia slender (3–4 times as long as broad) and the tibial spur is less than half the length of the tibia (c); prosternal process flat or arched or with a mid-line ridge at its widest point . . 15	

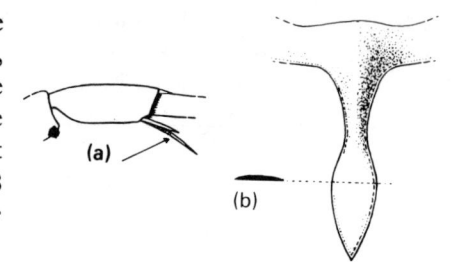

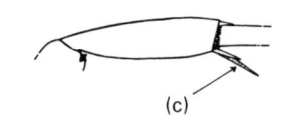

Note: Beware pale, soft immature specimens; if beetle is all brown and 9 mm long, but soft, rely on prosternal process and tibia characters.

15 (14)	Hind tarsal claws equal in length or nearly so (move swimming hairs aside) (a); sizes 6.0–9.0 mm 16	
—	Hind tarsal claws markedly unequal in length (b); sizes 8.5–11.5 mm . . . 23	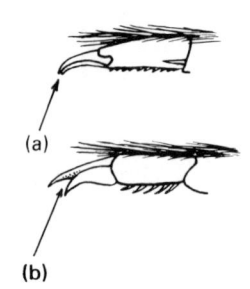
16 (15)	Pronotum with the row of punctures along the front edge with a gap in the centre (a) (light from the front and side, **dry** beetle); elytra usually with paler spots at the side margin; prosternal process with a low arch in cross-section at its **widest** point; size c. 9.0 mm 17	
—	Pronotum with a complete row of punctures along the front edge (b); elytra with or without paler marks at the sides; prosternal process with a high arch or a ridge at its widest point; sizes 6.0–8.5 mm . 18	

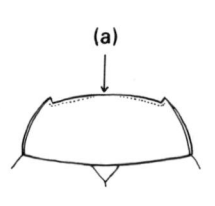

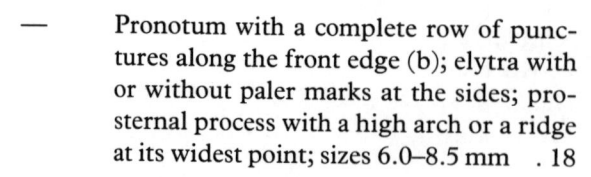

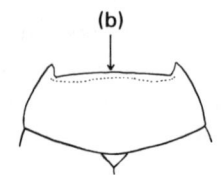

Note: A melanarius may come through to here if the slightly elongate reticulation is missed (see cpt 2); this species has a low-arch prosternal process but a complete row of punctures on the pronotum.

17 (16)	Last segment of the antenna with a dark tip (a); male with a tooth on the inner claw of the front tarsus (b) . . . . **Agabus biguttatus**	
—	Last segment of the antenna all brown; male without a tooth on the front tarsal claw . . . . . . **Agabus guttatus**	
18 (16)	Pronotum edged by a paler raised rim about as wide as the narrow part of an antennal segment (a); male with short broad front tarsal claws, the inner with a wide, blunt tooth (b); females shiny or dull; [size 7.0 mm] . . **A. uliginosus**	
—	Pronotum uniformly dark; lateral rim very thin, much narrower than an antennal segment; male front tarsal claws with a sharp tooth (c) or with no tooth 19	
19 (18)	Size usually c. 8.0 mm, but sometimes as small as 6.0 mm; black, brassy beetles; males without a tooth on the front tarsal claw (see note on sexing beetles at top of key) 20	
—	Size 6.0–6.5 mm; either all black, in which case males have a tooth on the front tarsal claw, or all brown, in which case males do not have a tooth on this claw 21	

Note: A mature, reddish beetle 7–8 mm long may be a variant of *A. undulatus*. Compare the underside with the figures in couplet 23; *A. undulatus* resembles Fig (a), other *Agabus* spp. keying out here resemble Fig (c).

20	Male genitalia (see note on extracting
(19)	genitalia, p. 5) with sucker hairs on the tips of the parameres (a) . . . .
	Agabus chalconatus

| — | Male genitalia without sucker hairs on the tip of the parameres (b) . . . . . . |
| | ***Agabus melanocornis*** |

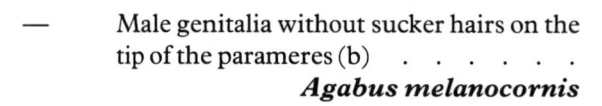

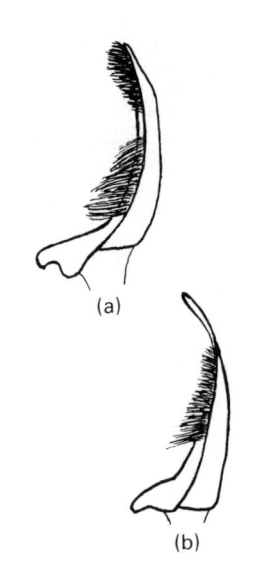

Notes: (1) **most** specimens with only the last antennal segment dark will be *chalconatus;* specimens with most segments dark can be **either** *melanocornis* or dark *chalconatus*. The male genitalia are the only criteria for certain identification.
(2) small *Ilybius* may key through to here; males and females of these *Ilybius* have a ridge on the last abdominal segment (see cpts 25 and 26).

| 21 | Hind margin of the metasternum moder- |
| (19) | ately curved, with low arches, not reaching the mid coxae (a); hind legs short (when stretched out behind only last 1/3 of the tibia overreaching the end of the elytra); males with a tooth on the front tarsal claw and a "file" of radiating ridges on either side of the third visible abdominal segment (b); black beetles . 22 |

| — | Hind margin of the metasternum with high arches, almost level with the hind edges of the middle coxae (c); hind legs long, (stretched-out tibia far outreaching the end of elytra); males without a tooth on the front tarsal claw, and no "file"; bronzy dark brown beetles; [size 6.0 mm] . . . |
| | ***Agabus labiatus*** |

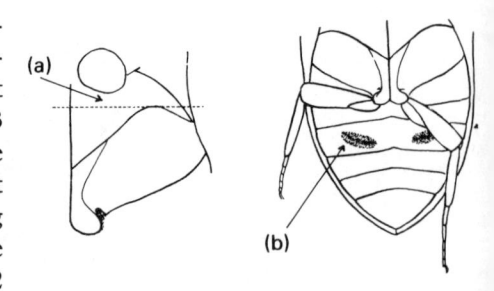

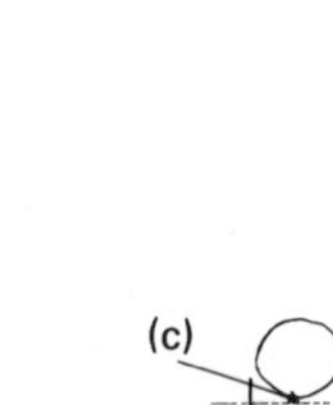

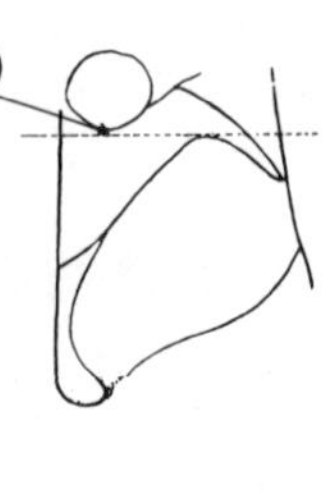

22 (21)	Last 6 antennal segments darkened on the end half (a); male with the ridges of the file of approximately equal length (b); male front tarsal claws with a sharp tooth (c); [size 6.0–6.5 mm] ***Agabus unguicularis***	
—	End third of the last antennal segment darkened, with the tip (at most) of other segments dark (d); ridges of the male file are of varying lengths (e); male front tarsal claws with a blunter tooth (f); [size 6.0–6.5 mm] . . . . . ***Agabus affinis***	
23 (15)	Lateral extremities of the metasternum very narrow (a); notch for the prosternal process not reaching the hind margin of the middle coxae (b); colour dark red or bronze; [size 11.5 mm] . . . . ***Ilybius fenestratus***	
—	Lateral extremities of the metasternum wide (c); notch for the prosternal process reaching the hind margin of the middle coxae (d); colour black or dark bronze 24	
24 (23)	Underside dark red, no ridge on the last abdominal segment (underside) of males or females (see note ii at top of Key); [size 11.0 mm] . . . ***Ilybius subaeneus***	
—	Underside black; males with a ridge on the last abdominal segment (view on a dry beetle, light from the side) 25	

25	Size 10.5–11.5 mm; last abdominal segment of the male with a long ridge extending about 1/3 the way to the front edge (a); the female with a ridge and a shallow indentation (b) . . . .
(24)	***Ilybius quadriguttatus***

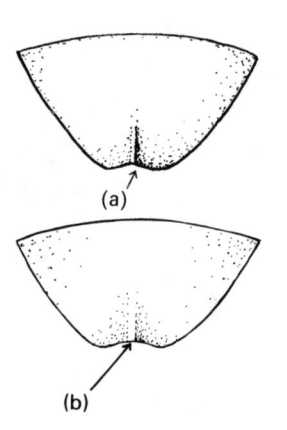

—	Size 10.0 mm or less; last abdominal segment of both sexes with a short ridge, extending at most 1/8 the way to the front edge 26

26	Beetle black with faint metallic reflections; males usually with two ridges on the midline of the hind coxal process (a); tip of the penis (see note on extraction of genitalia, p. 5) symmetrically pointed in side view (b); females with a small indentation in the tip of the abdomen and only a slight ridge (c); [size c. 9.0 mm] . . . . . .
(25)	***Ilybius aenescens***

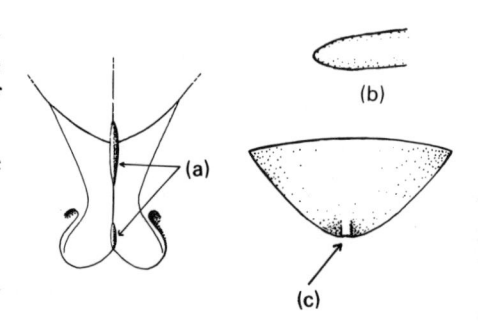

—	Beetle black with no trace of metallic reflections; males usually with only one ridge on the hind coxal process (d), and the tip of the penis with a profile like the prow of a ship (e); females with the tip of the abdomen scooped out on either side of a prominent knob (f); [size 9.5–10.0 mm] .
	Ilybius guttiger

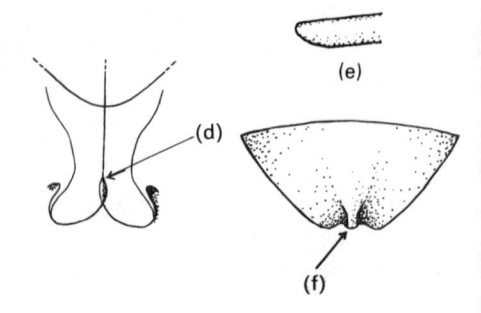

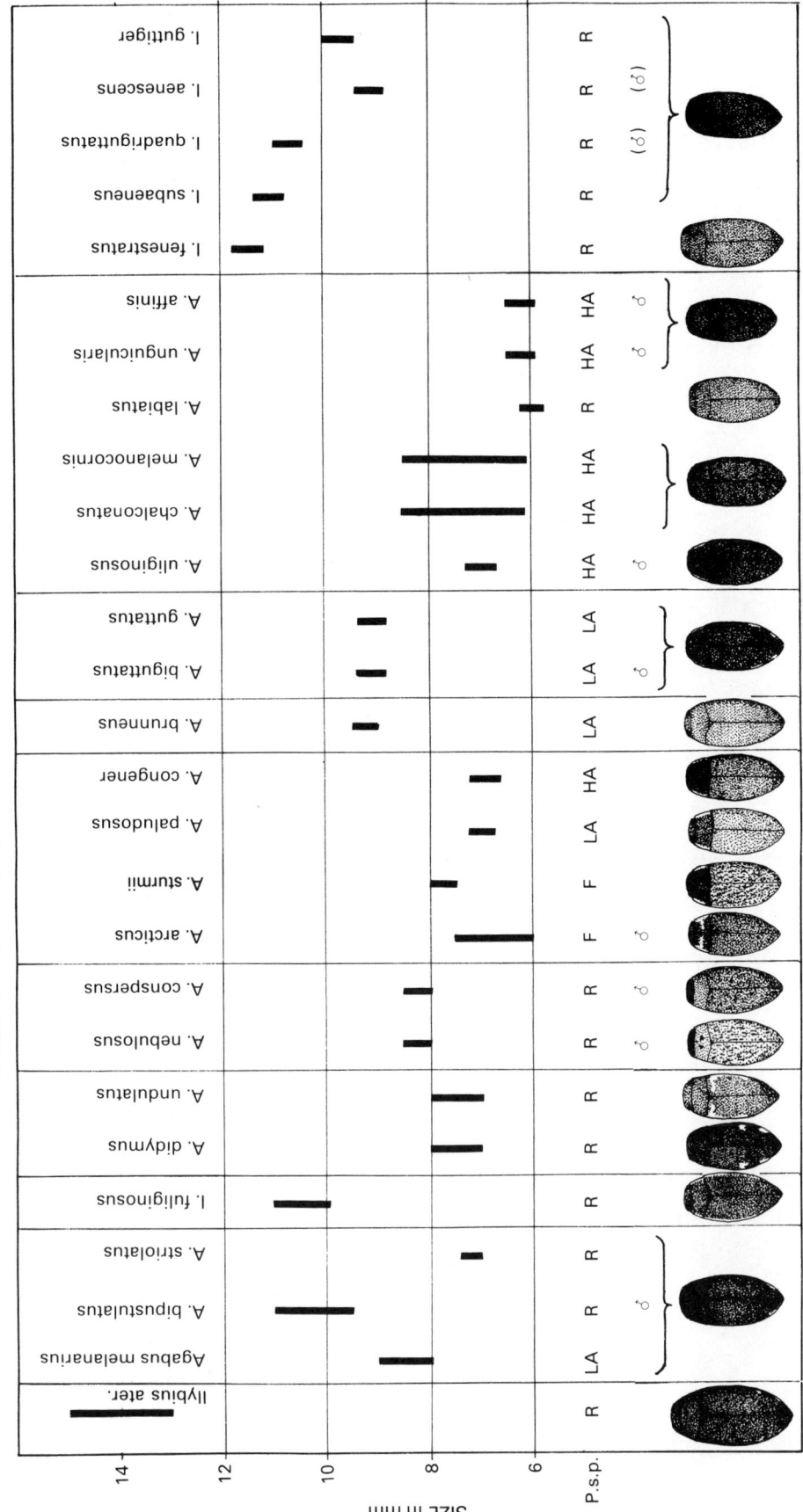

Agabus and *Ilybius* species (in order in which they appear in the key) size, prosternal process (psp) cross-sectional shape, male claws, and colour patterns

Figure 8. Tabular guide to *Agabus* and *Ilybius* species. To be used in conjunction with Key D1b.

KEY Prosternal process (psp) cross-section: F = flat, LA = low arch, HA = high arch/ridge, R = ridge; male characters: ♂ = front tarsal claw with a tooth or lobe; (♂) = in some cases not very obvious

Key D2   **Dytiscinae**

Sub-family Characters: large size; normally no indentation on the front of the eye; males with the middle segments of their front tarsi expanded into round plates. Size range 12.0–38.0 mm.
4 genera.

Note: Table 1 (p. 142) is an additional guide based on the colours of Colymbetinae and Dytiscinae species.

1 Size 22.0 mm or more; hind claws equal in length . . . . ***Dytiscus***
Key D2a, p. 51

— Size 18.0 mm or less; hind claws unequal
2

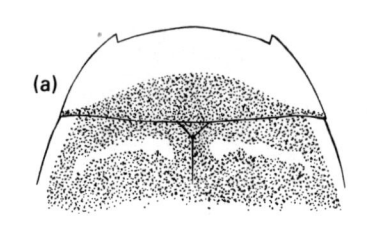

2 Black crescent on hind margin of the pronotum (a); elytra black with pale lateral margins . . . .***Hydaticus***
Key D2b, p. 53

— Black markings on the pronotum in bars (b); elytra mottled (use × 10 or more) . 3

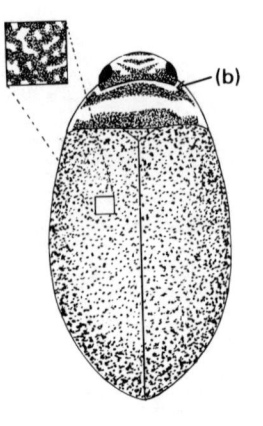

3   Pronotum with wide yellow borders, and a black bar across centre enclosing a bar of yellow (a); swimming hairs of the hind tarsi very long; females with wide hairy grooves on the elytra . . . . ***Acilius***
Key D2c, p. 53

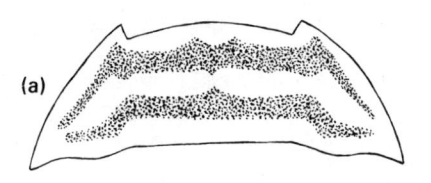

—   Black bands on the front and hind margins of the pronotum (a thin yellow line may be present on front margin) (b); swimming hairs short; neither males nor females with grooves on the elytra . ***Graphoderus***
Key D2d, p. 54

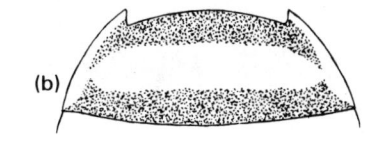

Key D2a   ***Dytiscus***

Genus Characters: large size; hind claws equal; female may have 9–10 deep grooves in elytra. Size range 22.0–38.0 mm.
6 spp.

1   Pronotum with yellow margin all round (a); hind coxal process with a sharp point (see couplet 2, Figs a & b) 2

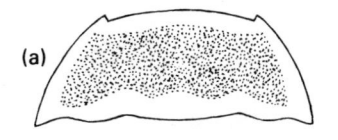

—   Pronotum with yellow on lateral borders only (b) (and possibly on anterior edge); hind coxal process without a sharp point (see couplet 5, Figs a & b) 5

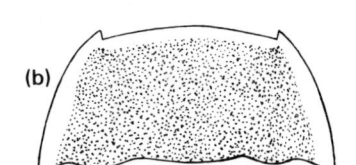

Note: the only other British water beetle larger than 22 mm is *Hydrophilus* (Key G, p. 93), which has no yellow borders on the pronotum. Recheck couplet 2 in the Family Key, p. 12.

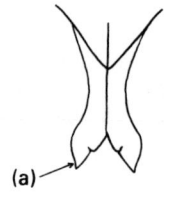

2 (1)	Hind coxal process with a short point (a); abdomen uniform yellow-brown below [size 26.0–32.0 mm] . **D. marginalis**	
—	Hind coxal process with a long point (b)	3

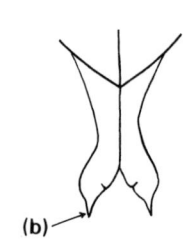

3 (2)	Abdominal segments uniform yellow-brown below [size 27.0–32.0 mm] . . . **D. circumcinctus**	
—	Abdominal segments with black borders	4

4 (3)	Elytra metallic green; abdomen predominantly green below; head uniformly dark [size 26.0–32.0 mm]. **D. circumflexus**
—	Elytra brown with yellow flecks; abdomen predominantly yellow-brown; yellow triangle between eyes [size 22.0–28.0 mm] . **D. lapponicus**

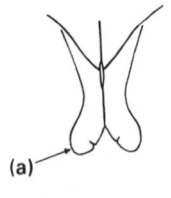

5 (1)	Abdomen uniformly black or dark brown; hind coxal lobes smoothly rounded (a) [size 22.0–30.0 mm] **D. semisulcatus**
—	Abdomen uniformly yellowish brown; hind coxal lobes elongated to a "corner" (b) [size 32.0–38.0 mm] **D. dimidatus**

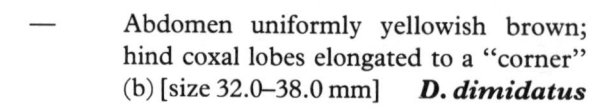

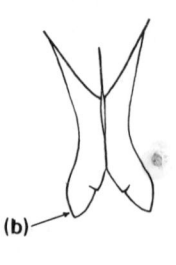

Key D2b   *Hydaticus*

Genus Characters: elytra black with pale lateral margins; black crescent on pronotum.
Size range 12.0–14.5 mm.
2 spp.

1   Elytra with transverse pale bars on the shoulders; black crescent extending forward about half way across the pronotum; (a) [size 12.0–13.0 mm] **H. transversalis**

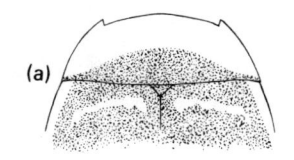

—   Elytra without bars on the shoulders; black crescent extending 3/4 way across the pronotum (b); [size 13.0–14.5 mm] . **H. seminiger**

Note: *H. stagnalis*, which has a narrow crescent on the pronotum but no pale bars on the elytra, is now probably extinct in Britain.

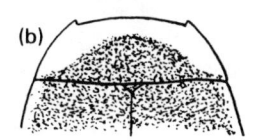

Key D2c   *Acilius*

Genus Characters: elytra mottled; black bar on pronotum; form "pear-shaped"; females with hairs in grooves on elytra.
Size range 14.0–18.0 mm.
2 spp.

1   Bases of the hind femora black; female with 2 patches of hair on the pronotum (a) [size 16.0–18.0 mm]. . . **A. sulcatus**

—   Hind femora uniformly pale; females without patches of hair on the pronotum [size 14.0–16.0 mm]. **A. canaliculatus**

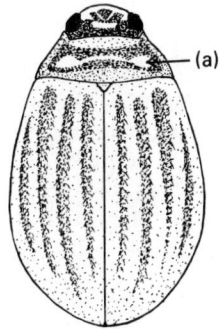

Key D2d *Graphoderus*

Genus Characters: black bands on front and hind margins of pronotum; elytra mottled.
Size range 13.0–16.0 mm.
3 spp.

Note: all 3 spp are very rare and require checking by experts.

1 Black band on the hind margin of the pronotum narrow, less than half the width of the central yellow band (a); body pear-shaped . . . . **G. bilineatus**

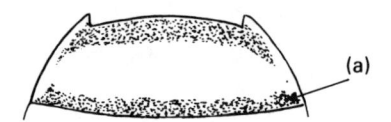

— Black band on the hind margin of the pronotum about as wide as the central yellow band; body oval 2

2 (1) Narrow yellow band present between the front black band and the anterior edge of the pronotum, at least behind the eyes (centre may be darkened) (a); male with sucker hairs on the **middle** tarsi in two irregular rows, sometimes more than two abreast; female with the shorter **hind** tarsal claw at most half as long as the longer one (b) . . . . **G. zonatus**

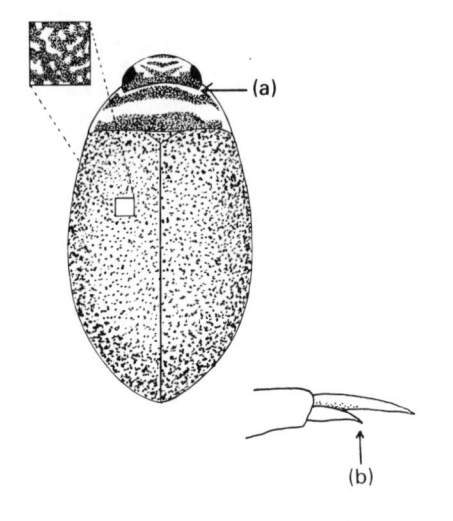

— No yellow band on the front margin of the pronotum (c); male with sucker hairs on the **middle** tarsus in two regular rows (d); female with the shorter **hind** tarsal claw more than half as long as the longer one (e) **G. cinereus**

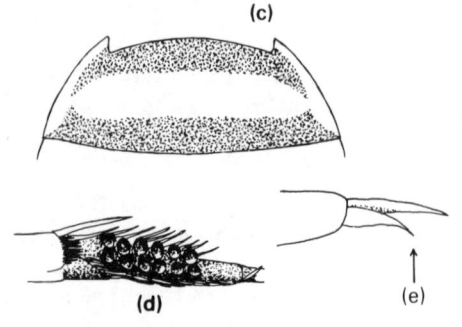

Key D3 Laccophilinae

Sub-family Characters: no mesoscutellum visible; lobed hind tarsal segments.
Size range 3.0–4.5 mm.
1 genus Laccophilus *3 spp.*

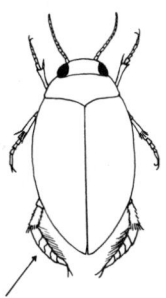

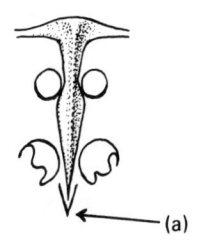

1 Size c. 3.0 mm; elytra dark brown/black with yellow in patches; prosternal process needle-like, the groove in which it lies reaching beyond the hind margin of the mid coxae (a) . . . . . ***L. obsoletus***

— Size 4.0–4.5 mm; elytra brown or greenish, with or without paler flecks; prosternal process short and broad, its groove not reaching beyond the mid coxae (b) . . 2

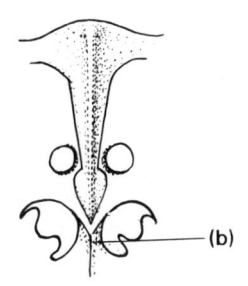

2 Group of radiating ridges (a) (stridulatory file) present on the hind coxae of both sexes . . . . ***L. hyalinus***
(1)

Note: Look **in front** of the hind coxal lobes, moving legs aside if necessary.

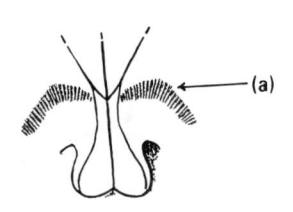

— No stridulatory file in either sex . . . . ***L. minutus***

Key D4   **Hydroporinae**

Sub-family Characters: no mesoscutellum; front and middle tarsi apparently 4-segmented, the third deeply lobed.
Size range 1.5–6.0 mm.
17 genera.

Note: Table 2 (p. 144) is an additional guide based on the colour patterns found in the Hydroporinae.

1   Elytra black with 12 yellow spots, (a); pronotum bulging out at sides (b); size 4.5–6.0 mm 2

—   Size and coloration not as above; pronotum bulging or (more usually) evenly tapering from the rear to the front; size 1.5–5.5 mm 3

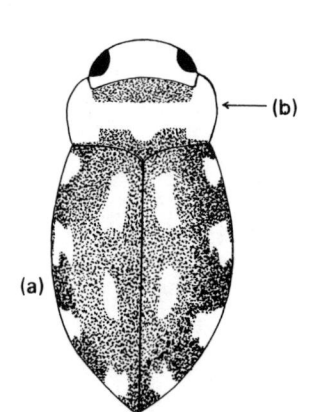

2   Each elytron with a small tooth projecting
(1)   from the margin near the tip (a) (most easily seen from below at × 12); elytra pattern usually with longitudinal lines as well as spots (b) . . . . ***Potamonectes***
                                             Key D4g, p. 84

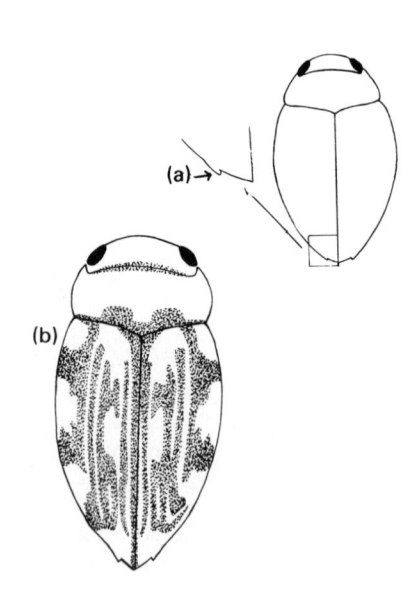

—   No teeth on the elytra; elytra pattern without longitudinal lines (couplet 1, Fig (a)); [size 5.0–6.0 mm] . . . .
Stictotarsus duodecimpustulatus

A Key to the Adults of British Water Beetles

3 (1)   Pronotum with two longitudinal furrows placed 2/3 way from the mid-line to the lateral margin **and** matched by a pair of furrows in the elytra (a); size 1.5–2.0 mm.
             Bidessus* and *Hydroglyphus
             Key D4c, p. 66

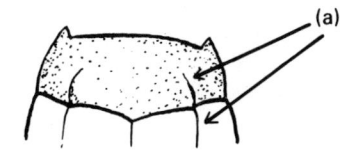

—   Pronotum without furrows in this position; if present, furrows are nearer the lateral edge (b); furrows in the elytra, if present, do not match those on the pronotum 4

Note: Longitudinal furrows are marked, at least on their outer edge, by a sharp vertical edge, which is visible as a dark line when **illuminated from the front or side.** Distinguish such an edge from a pigment line and from the raised rim on the extreme edge of the pronotum. Furrows, where present, are at least as obvious as the rim. **View at ×20 on a dry beetle.**

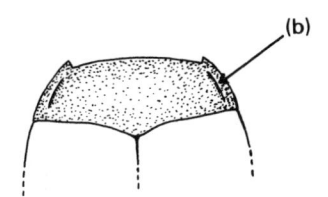

4 (3)   Body shape, viewed from above, a very broad oval, with the elytra coming to a point at the tip (a); antennae with cup-shaped segments which are wider than long (b); prosternal process (between the front coxae) fan-shaped and concave in cross section (c); [size 2.0–2.5 mm] . . .
             Hydrovatus clypealis

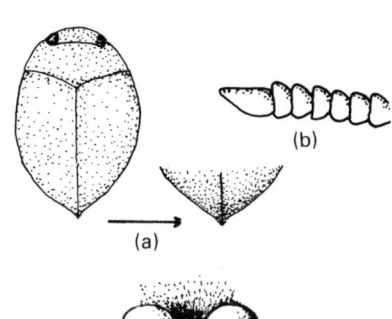

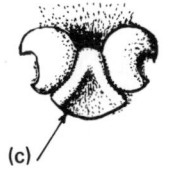

—   Body shape rounded or more elongate, the elytra not coming to a point at the tip; antennae with all segments longer than wide; prosternal process oval (pointed or rounded) at the tip (d) size 2.0–5.5 mm   5

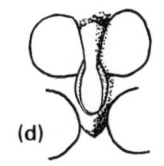

5 (4)	Body globose, very convex below (a), total length of the beetle only about twice to $2\frac{1}{4}$ times its maximum depth, seen from the side; seen from above, body shape a rounded oval (b); pronotum never with furrows. 6	
—	Body only moderately convex above and below (c), its total length about $2\frac{1}{2}$ to 3 times its maximum depth; seen from above, body shape either an elongate oval (d) or, in some species less than 3 mm long, a rounded oval; pronotum with or without furrows 7	
6 (5)	Size 4.5–5.0 mm; elytra chestnut, with or without slight mottling (or, in Channel Isles, with distinctive black and yellow pattern—see Key D4a); hind tarsal claws very unequal in length . . ***Hyphydrus*** Key D4a, p. 62	
—	Size 2.0–4.0 mm; elytra with distinctive colour patterns: either longitudinal lines (a) or blotches (b); hind tarsal claws equal in length . . . . ***Hygrotus*** and ***Coelambus*** (part) Key D4b, p. 63	

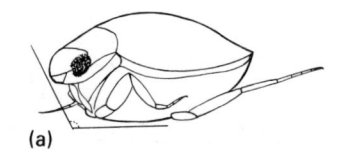

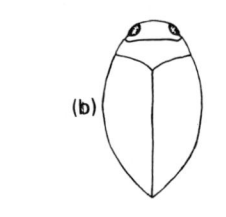

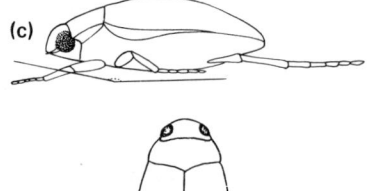

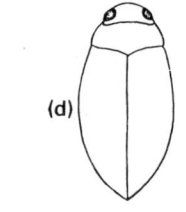

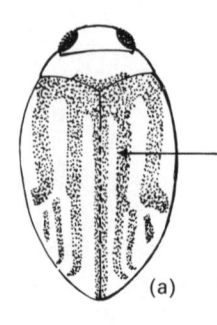

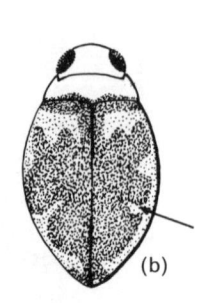

7	End of the prosternal process rounded, the end part flat in cross-section with a raised border (a) (×20, dry beetle, light from the side); body elongate, parallel-sided (b); colour chestnut with the pronotum black, darker than the head and elytra [size 5.0 mm] . . . .
(5)	***Laccornis oblongus***

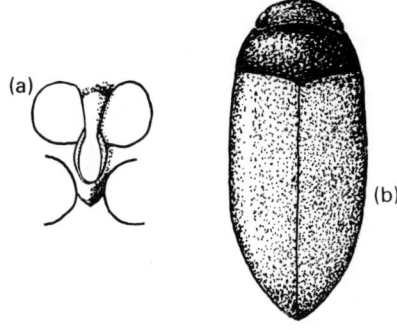

— End of the prosternal process more pointed, and arched or with a mid-line ridge in cross-section and without a raised border (c); size 2.0–5.5, (if as large as 5.0 mm, elytra well rounded at the sides and pronotum not uniformly black) . 8

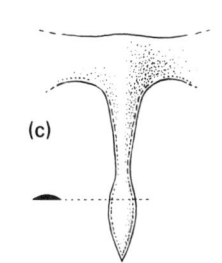

8	Lines on the hind coxal process running parallel and close together towards the anterior end (a) raised into obvious ridges half way along in males (b); body shape broad; elytra red-brown with paler shoulders (c), with dense fine punctures and large shallow pits; [size 4.5 mm] . .
(7)	***Deronectes latus***

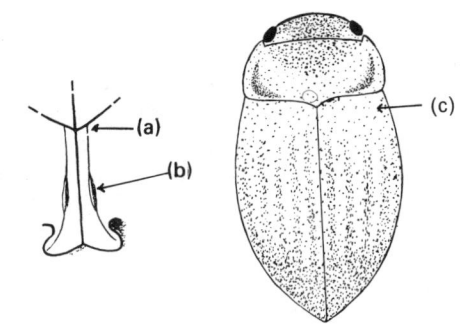

— Lines on the hind coxal process diverging as they run forwards (d) (and may converge again); beetle not as above. . . 9

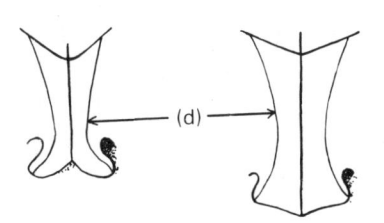

9 (8)	Pronotum with a longitudinal furrow on each side (a) (view on a dry beetle at × 15 or more with the light from the front and side; furrows are marked on the outside by a definite edge and are at least as obvious as any raised rim on the side of the pronotum); hind coxal process with rounded lobes (b) 10	
—	Pronotum without deep furrows, but may have shallow, wide depressions (without a marked edge) near the side or hind margins; hind coxal process either with lobes as above, or with a straight (c) or wavy (d) hind margin 11	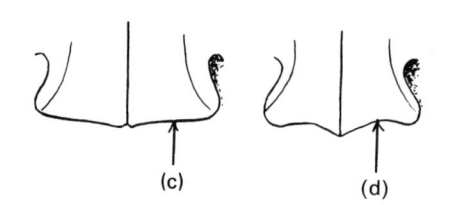
10 (9)	Elytra with dark anterior margins; pronotum dark with, at most, paler borders (a) (b); no furrows in the elytra; size 2.0–3.0 mm . . . . ***Graptodytes* and *Stictonectes*** Key D4d, p. 67	
—	Elytra with pale anterior margins; pronotum predominantly pale, usually with darker blotches in the centre; elytra each with 3 faint furrows (c); size 3.0–4.5 mm . ***Oreodytes*** Key D4e, p. 68	

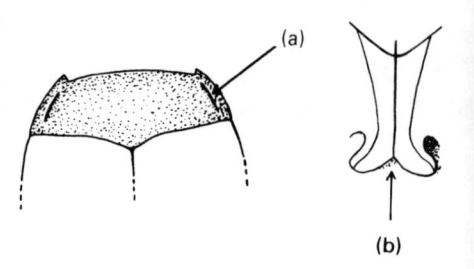

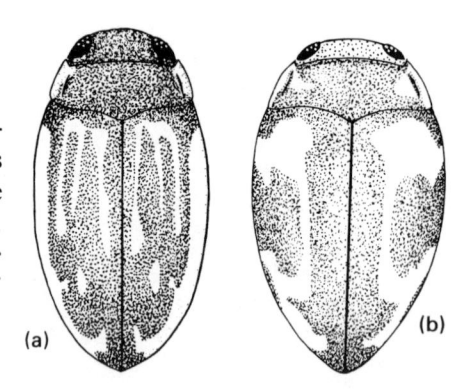

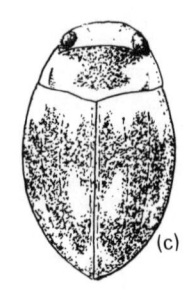

11	Hind coxal process with rounded lobes (a); elytra usually with distinct longitudinal lines of colour (yellow and black, or chestnut and dark brown), though the lines may be faint or obscured by blotches; size 3.0–5.2 mm 12
(9)	

Note: Allow alcohol-preserved specimens to dry until full colours develop.

— Hind coxal process with a straight (b) or wavy (c) hind margin, sometimes with slight lobes; elytra never with lines of colour, though pale blotches may be present; size 1.7–5.5 mm . . . . . .
Hydroporus and **Suphrodytes**
Key D4f, p. 70

12	Underside of the thorax and abdomen without large punctures, but with a dense covering of fine punctures giving an appearance rather like sandpaper (a) . .
(11)	

Potamonectes
Key D4g, p. 84

— Underside with large punctures, with a shining (b) or matt (c) surface between . 13

13	Size 4.0–5.0 mm; underside of the thorax and abdomen black; elytra widest 1/2 to 3/4 the way back (a). 14
(12)	

— Size 3.0 mm; underside of the thorax and abdomen uniformly red or yellow; elytra chestnut with nine darker lines; elytra widest about 1/4 the way back and tapering to an point (b); densely hairy . . .
Porhydrus lineatus

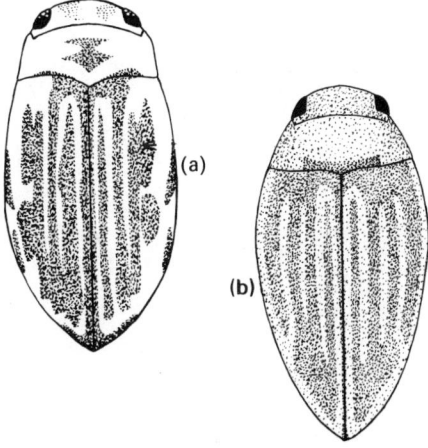

14 (13)	Underside smooth and brightly shining between punctures (a); elytra yellow with black blotches and lines (b); front and middle tarsi entirely black; [size 4.0 mm]. ***Scarodytes halensis*** Key D4e, p. 68	
—	Underside with net-like reticulation between punctures (view at ×20 or more on a dry beetle) (c); front and middle tarsi all pale or with only the last two segments darkened (d); size 4.0–5.0 mm . . . . ***Coelambus*** (part) Key D4b, p. 63	

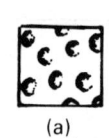

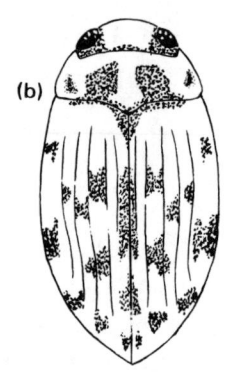

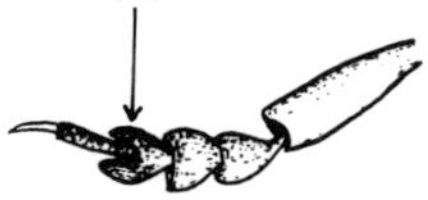

Key D4a   *Hyphydrus*

Genus Characters: globose beetles.
            Size range 4.5–5.0 mm.
            2 spp.

1	Colour red, with or without mottling . . **H. ovatus**
—	Colour yellow-red with black markings (a) **H. aubei** (Channel Islands)

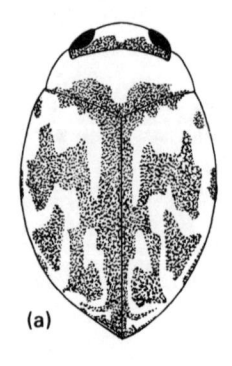

A Key to the Adults of British Water Beetles 63

Key D4b *Hygrotus* and *Coelambus*

Genera Characters: no furrows in pronotum.
Size range 2.0–5.0 mm.
Hygrotus: *4 spp.*
Coelambus: *5 spp.*

Note: Puncturation of elytra is most obvious at ×20 and above, lit from the side on a **dry** beetle.

1	Body globose (a), strongly convex below; underside smooth and brightly shining between large punctures; size 2.0–4.0 mm 2	
—	Body elongate (b), only slightly convex below; underside with reticulation between large punctures; size 4.0–5.0 mm . . 7	
2 (1)	Abdomen black, strongly contrasting with the yellow borders of the elytra seen from below (the epipleurs) (a) 3	
—	Underside uniformly red or yellow . . 4	
3 (2)	Black lines on the elytra beginning just behind the front margin; [size 3.5–4.0 mm] . ***Coelambus nigrolineatus***	
—	Black lines beginning 1/3 way down the elytra, leaving the "shoulders" pale (a); [size 3.0–3.5 mm] . . . . ***Coelambus confluens***	

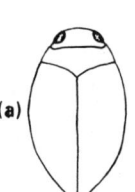

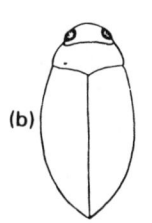

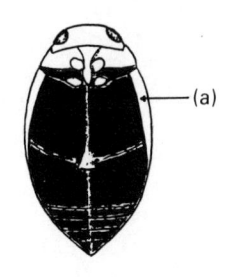

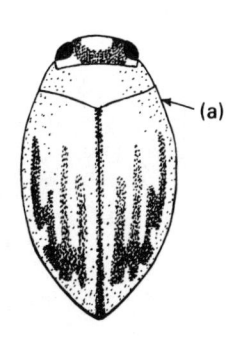

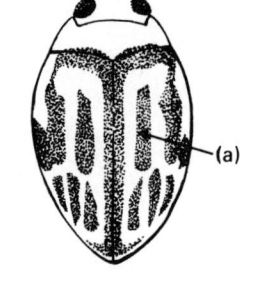

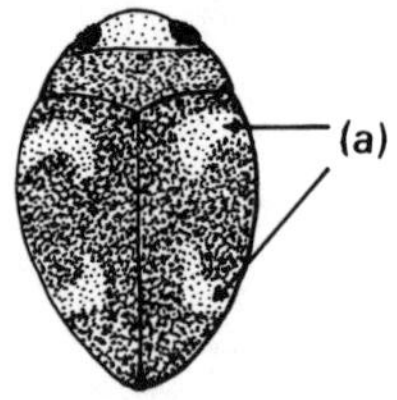

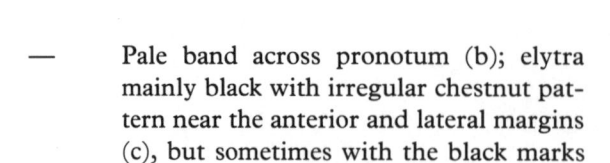

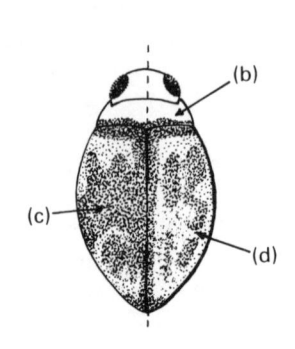

4 (2)	Size 4.0 mm; elytra predominantly pale with longitudinal black lines 5	
—	Size 2.0–3.0 mm; elytra usually predominantly dark with irregular paler patches but sometimes predominantly pale with darker blotches or lines 6	
5 (4)	Large punctures on the elytra sparse and far outnumbered by small ones (×30); dark lines nearest the centre of the elytra usually broken into bars (a); [size c. 4.0 mm] . . . ***Hygrotus versicolor***	
—	Large punctures on the elytra as numerous as small ones; dark lines nearest the centre of the elytra usually running full length (b); [size c. 4.0 mm] . . . . . ***Hygrotus quinquelineatus***	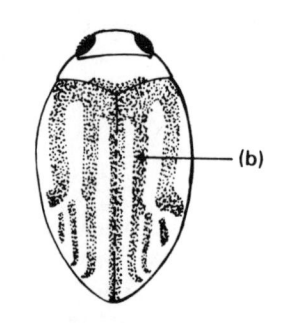
6 (4)	No pale band across pronotum; elytra chestnut with paler patches near the anterior margin and the tip (a); [size 2.0 mm] . . . ***Hygrotus decoratus*** **Note:** *Graptodytes pictus* (Key D4d, p. 67) has similar coloration, but is less convex and has pronotal furrows.	
—	Pale band across pronotum (b); elytra mainly black with irregular chestnut pattern near the anterior and lateral margins (c), but sometimes with the black marks broken up into bars (d); [size 3.0 mm] . . ***Hygrotus inaequalis***	

| 7 | Elytra each with 3 distinct longitudinal furrows formed of coalesced large punctures (×20) (a); elytra chestnut with darker stripes; tarsi yellow; [size 5.0 mm] . . . . |
| (1) | ***Coelambus impressopunctatus*** |

— Elytral furrows very faint; colour yellow with black stripes; tarsi yellow, with or without dark markings (see below) . . 8

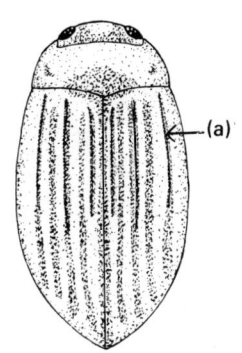

| 8 | Legs and antennae yellow or red with dark tips on the last few segments of the tarsi (a) and antennae; second pair of dark stripes from the centre of elytra usually do not meet the dark patch on the anterior margin of the elytra (b); [size 4.0 mm] . . . . |
| (7) | ***Coelambus novemlineatus*** |

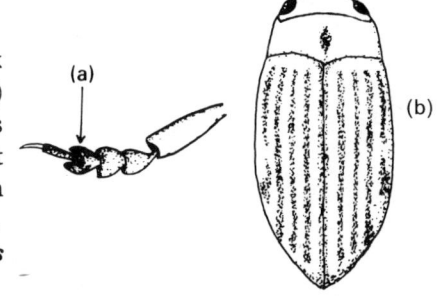

— Legs and antennae red, with no dark marks on the tarsi; second pair of stripes from the centre joining with the dark pattern on the anterior margin of the elytra (c); [size 5.0 mm] . . . .
Coelambus parallelogrammus

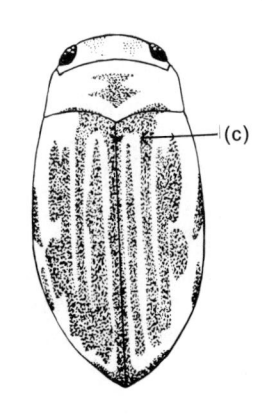

Key D4c *Bidessus* and *Hydroglyphus*

Genera Characters: furrows on the pronotum matched by those on the elytra.
Size range 1.5–2.0 mm.
Bidessus: 2 spp.
Hydroglyphus: *1 sp.*

1   Furrows on the elytra which match those on the pronotum extend about half way down the elytra (a); elytra elongate and parallel-sided, yellow with brown bars across [size 1.5 mm]. . . . .
Bidessus minutissimus

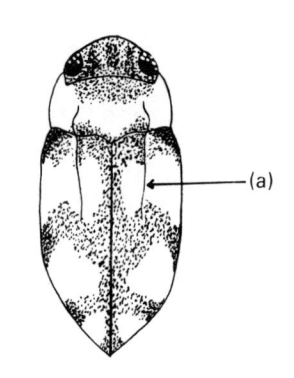

—   Furrows on the elytra which match those on the pronotum short, not extending more than 1/3 down the elytra; body oval
2

2   Elytra brown; a pair of furrows running parallel and close to the midline are indistinct and do not extend more than half way down the elytra (a) [size 2.0 mm] . . .
(1)
Bidessus unistriatus

—   Elytra dark brown with yellow markings; furrows running parallel and close to the midline are distinct and run almost to the tip of the elytra (b) [size c. 2.0 mm]. . .
Hydroglyphus pusillus

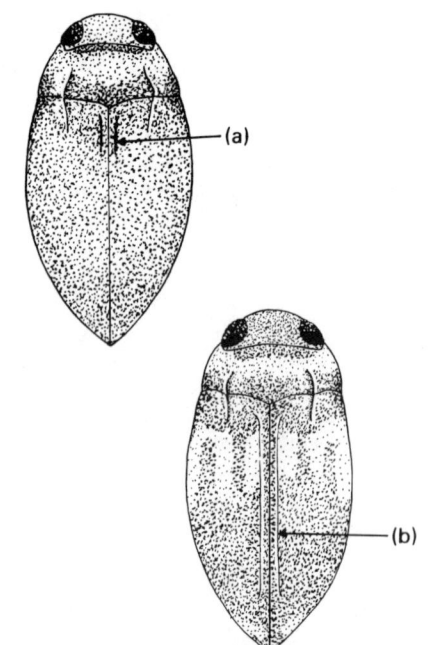

A Key to the Adults of British Water Beetles

Key D4d *Graptodytes* and *Stictonectes*

Genera Characters: furrow near lateral margin of pronotum; elytra predominantly dark with paler patterning; hind coxal process with rounded lobes.
Size range 2.0–3.0 mm.
Graptodytes: *4 spp.*
Stictonectes: *1 sp.*

1   Elytra with paler colour in transverse blotches; body shape a rounded oval . 2

—   Elytra with paler colour in longitudinal lines; body shape an elongate oval . . 3

2 (1)   Yellow zig-zag bands across the elytra (a); centre of the pronotum black; [size 3.0 mm] . . . ***Stictonectes lepidus***

—   Coloration as shown (b); head red; [size 2.0–2.4 mm] . . ***Graptodytes pictus***

Note: *Hygrotus decoratus* (Key D4b) has similar coloration, but has no pronotal furrows and is very convex below.

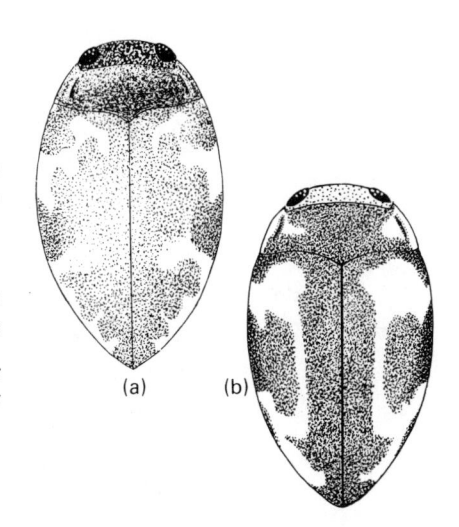

3 (1)   Elytra each with about 5 yellow lines (a); [size 2.5 mm] . .***Graptodytes flavipes***

—   Elytra each with, at most, 2 lighter lines visible (b) 4

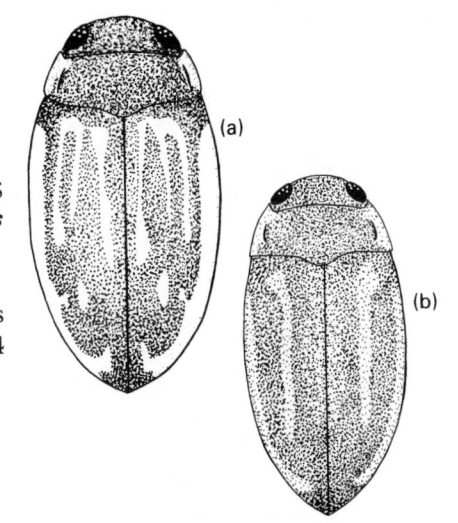

4 (3)	Antennal segments 5–9 slender, distinctly longer than broad (a); body almost parallel-sided; male front tarsal claws very unequal in length (b); [size 2.3–2.7 mm] . ***Graptodytes bilineatus***	
—	Antennal segments 5–9 hardly longer than width across their end margin (c); body more rounded; male front tarsal claws equal in length (d); [size 2.0–2.3 mm] . . ***Graptodytes granularis***	

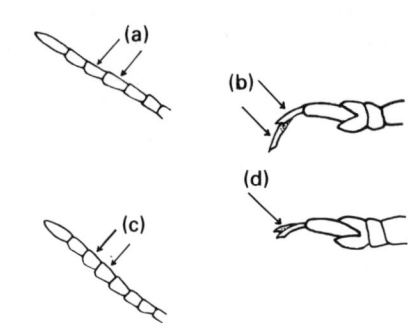

Note: If uncertain of your identification, and if the beetle is more than 3.0 mm long, check Key D4e.

Key D4e   *Oreodytes* and *Scarodytes*

Genera Characters: lateral furrows on the pronotum (not obvious in Scarodytes*); the elytra pattern variable but based on 6–7 dark lines on each elytron coalescing into spots; pronotum pale, usually with dark blotches.*
Size range 3.0–4.5 mm.
Oreodytes *4 spp.*
Scarodytes *1 sp.*

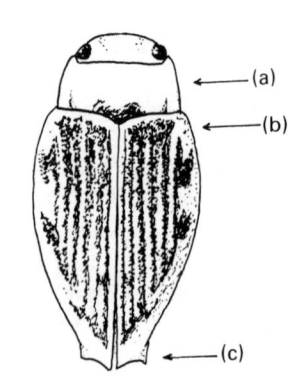

1	Pronotum almost parallel-sided, rather square (a); elytra very broad at shoulder (b); female with large tooth at the tip of each elytron (c) . .***Oreodytes alpinus***	
—	Pronotum with curved sides, narrowing markedly towards front; elytra less broad at shoulder (see following couplets); no teeth on elytra in either sex 2	

A Key to the Adults of British Water Beetles

2 (1)	Thorax (underside) and the hind coxae glossy and shining brightly between large punctures; furrows on the pronotum very indistinct and not visible on the elytra; coloration as shown (a); [size 4.0 mm] . . **Scarodytes halensis**	
—	Underside with granular appearance between punctures; pronotal furrows with sharp outer edge (b). 3	
3 (2)	Size 4.0 mm or more; lines on the hind coxal process turning inwards at the anterior margin (a); elytra with dense large punctures between those forming furrows (×20 or more, lit from side) . . . . . **Oreodytes davisii**	
—	Size 3.5 mm or less; lines on the hind coxal process continuing to diverge up to the anterior margin (b) 4	
4 (3)	Body shape long oval (a), abdomen always black below; elytra with large punctures scattered thickly between those forming furrows; [size 3.2–3.5 mm] . . . . . **Oreodytes septentrionalis**	
—	Body shape round oval (b); abdomen may be orange or black below; hardly any punctures on the elytra other than those forming furrows; [size 3.0 mm] . . . . **Oreodytes sanmarkii**	

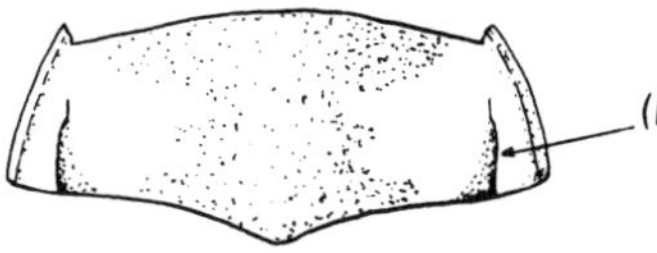

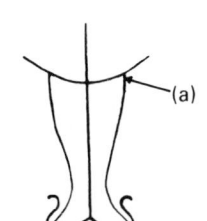

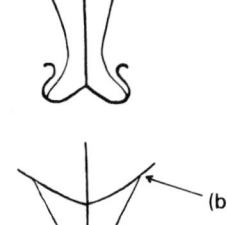

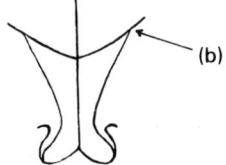

Note: If the beetle is less than 3.0 mm long, check Key D4d, p. 67.

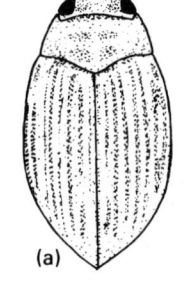

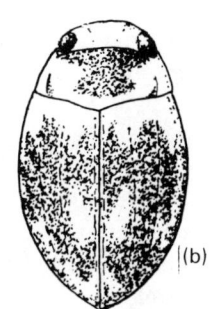

Key D4f   *Hydroporus* and *Suphrodytes*

Genus Characters: hind coxal process with straight, wavy or slightly lobed margins.
Size range 1.7–5.5 mm.
Hydroporus *28 spp.*
Suphrodytes *1 sp.*

READ THESE NOTES BEFORE PROCEEDING

Male genitalia

Many species of *Hydroporus* are not possible to identify with certainty except by examination of the male genitalia (see notes on dissection on p. 5). The aedeagus (penis) should be displayed convex side up (see diagram) and viewed slightly moistened at high magnification. Figures of the aedeagus of all the British species of *Hydroporus* and *Suphrodytes* are redrawn in Figure 9, p. 83, from Foster & Angus (1985). It should be noted that, since the aedeagus is strongly curved, there is some foreshortening and the structure will need to be tilted back and forth when examined under the microscope.

Hydroporus are difficult to sex, both sexes having sucker hairs on the front and middle tarsi and, in many species, little difference in the width of these tarsi or the development of their claws. However, in some species, marked with a symbol in the table at the end of the key, the males have a lobe or tooth on one of the front tarsal claws, visible at $\times 50$. Otherwise, dissection is the surest way of distinguishing males and females.

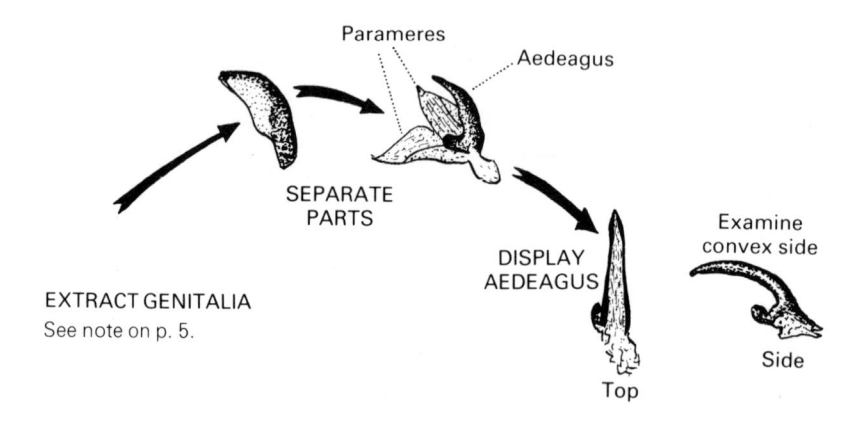

Reticulation on elytra

At couplet 5, the front quarter of the DRY beetle must be examined with the light from the side, at $\times 25$ or more. Where reticulation is present, the elytra have a slightly frosted appearance at $\times 12$, at $\times 25$ the net-like reticulation between the punctures is usually visible and at $\times 40$ and above, the individual meshes are very obvious and can be counted (as in couplet 32). Where there is no reticulation, the elytra are smooth and shiny, although dense hairs make this less obvious at low magnification.

Colours

Spirit-preserved specimens should be allowed to dry for 5–10 minutes, sufficient for the elytral colours to be revealed. Note that dry-stored specimens fade, as do very old beetles (these will also tend to have lost their hairs) and that variation in colours occur in nature. Also, newly-emerged specimens will not show characteristic coloration; these can be recognised by the ease with which the cuticle is dented, and are best identified by reference to mature specimens. If there is doubt about colours, the table at the end of the key allows comparison of sizes and genitalia between the different colour groups.

The prosternal process (a)
The neck of the process must be viewed on the underside of a clean dry beetle, looking between the front coxae with the beetle's head tilted up at an angle of 45° and strongly illuminated.

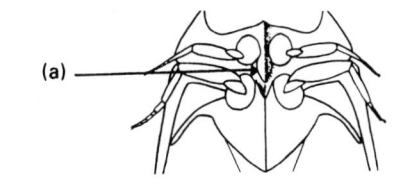

FIGURE 9 (p. 83) SHOWS THE SIZES AND MALE GENITALIA OF ALL THE BRITISH SPECIES AND SHOULD BE USED FOR CHECKING ALL IDENTIFICATIONS.

1 Size 4.5–5.5 mm; back of pronotum curving in sharply to meet the front of the elytra at a distinct angle when viewed from above; elytra black, usually with a distinctive yellow pattern of zig-zags (a) or shoulder spots (b) or may be almost completely black; pronotum with paler side margins; prosternal process flat at its broadest point and with a wide neck, at least half as wide as the width at the broadest point (c); aedeagus 1 . . . .
Suphrodytes dorsalis

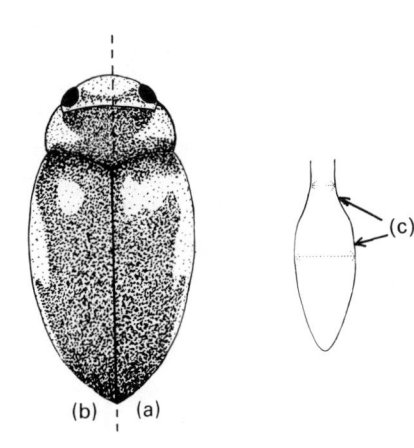

— Size mostly less than 4.7 mm (see Fig. 9, p. 83); angle between pronotum and elytra less obvious or absent; coloration not black and yellow, except in species less than 4.0 mm; prosternal process arched or with a midline ridge at the broadest point and a narrow neck, less than half the width at the broadest point 2

2 (1)	Elytra black with a distinctive yellow pattern as shown (a); [size 3.2–3.8 mm] . 31	
—	Elytra of other coloration and markings 3	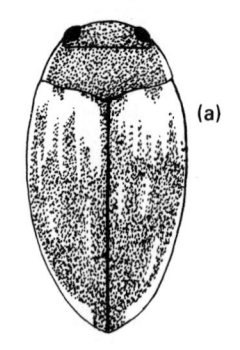
3 (2)	**Front quarter** of the elytra pale and transparent (a) so that trachea and wing folds may be visible through it; rest of elytra dark with pale flecks visible to the naked eye; head entirely black between the eyes (though its front may be paler) . 4	
—	Front quarter of elytra **opaque**, brown or black or with a paler narrow band along the front (the "shoulder"), or elytra **transparent all over** (see couplet 12); head red or black or with dark and light pattern between the eyes 5	
4 (3)	Size 3.5–4.5 mm; pronotum black with broad paler sides; elytra dull with dense hairs; hind coxal process with very dense hairs between the lines near the tip (a) (if these hairs have come off, the pits in which they originated will still be obvious at ×20); aedeagus 2 . . **H. marginatus**	
—	Size 3.4–3.8 mm; pronotum entirely black except perhaps for the extreme rim; elytra shiny with sparse hairs; hind coxal process with fewer hairs between the lines (b); aedeagus 3 . . . . **H. tessellatus**	

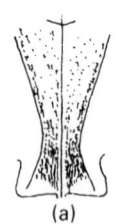

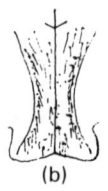

5 (3)	Front half of elytra without reticulation (see note above), shining between dense hairs; head completely black between the eyes, but may be paler at the front and hind edges. 6
—	Front half of elytra with reticulation, dull or shining, hairy or not hairy; head red, brown, black or patterned between the eyes 8
6 (5)	Last abdominal segment (underside) and centre of pronotum without reticulation, shining between punctures; [size 3.2–3.8 mm]; aedeagus 4 . **H. pubescens**
—	Last abdominal segment and pronotum with reticulation ($\times 40$), appearing dull at low magnification 7
7 (6)	Size 4.0–4.7 mm; elytra with pale shoulder markings; neck of prosternal process (see note, p. 71) without grooves; aedeagus 5 **H. planus**
—	Size 3.0–3.5 mm; elytra uniformly dark; neck of prosternal process with grooves (a); aedeagus 6 . . . . . **H. discretus**

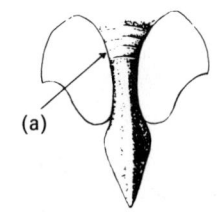

8	Head, pronotum and elytra all the same colour, either all red, all brown or all black	
(5)		9

— Centre of pronotum darker than the front of the elytra (the difference in shade can usually be detected even by the naked eye in daylight); head red or brown, or black with red markings; elytra red/brown, darkening towards rear, or brown or black with a paler shoulder mark and/or distinctive yellow pattern 24

Note: Allow spirit-preserved beetles to dry—see note, p. 70. If the pronotum is paler than elytra, follow the first lead.

9	Head, pronotum and elytra all red, red/brown or mid-brown; pronotum may be lighter than elytra; (head may have vague darker markings and elytra may be mottled).	
(8)		10

— Head, pronotum and elytra all black or very dark brown; head entirely black between the eyes; elytra without paler patterning 18

ALL RED, RED/BROWN, MID-BROWN SPECIES

10	Hind tibia narrowing abruptly half way to the junction with the femur (a); [size 2.8–3.3 mm]; aedeagus 7	
(9)		
	male *H. angustatus*	

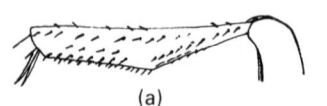

— Hind tibia narrowing gradually near junction with femur. 11

11 (10)	Size 2.2 mm or less; third segment of front tarsus very elongate, more than twice as long as second segment (×40) (a); male front tarsal inner claw with a lobe (×50) (b); aedaegus 8 . . . **H. scalesianus**	
—	Size 2.5 mm or more; third segment of front tarsus less than twice as long as the second segment (c); male front tarsal claw without a lobe (d) 12	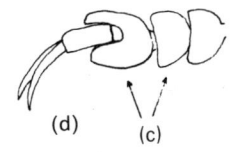
12 (11)	Elytra transparent so that wing folds may be visible along their entire length (beware soft specimens); elytra appear mottled at low magnification; elytra somewhat parallel-sided (a). 13	
—	Elytra opaque, of uniform coloration; elytra with smoothly-rounded outline (b) 14	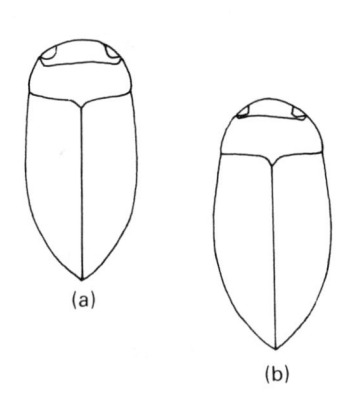
13 (12)	Pronotum with a group of very heavy elongate punctures on the hind margin near the sides and a row along the front edge, but virtually no large punctures anywhere else (a) (light from the side and in front, dry beetle); middle of elytra with sparse large punctures more than twice their own diameter apart and tiny punctures (visible at ×40) in between; prosternal process without grooves (b) (see note, p. 71 [size 3.3–4.2 mm]; aedeagus 9. . . **H. obsoletus**	
—	Pronotum with large punctures more widely distributed, though with a concentration at the hind margin and fewer in the centre (c); centre of elytra with large punctures less than twice their own diameter apart and no smaller punctures in between; prosternal process with grooves (d); [size 3.5–4.2 mm]; aedeagus 10. . . **H. ferrugineus**	

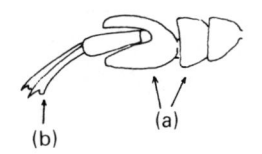

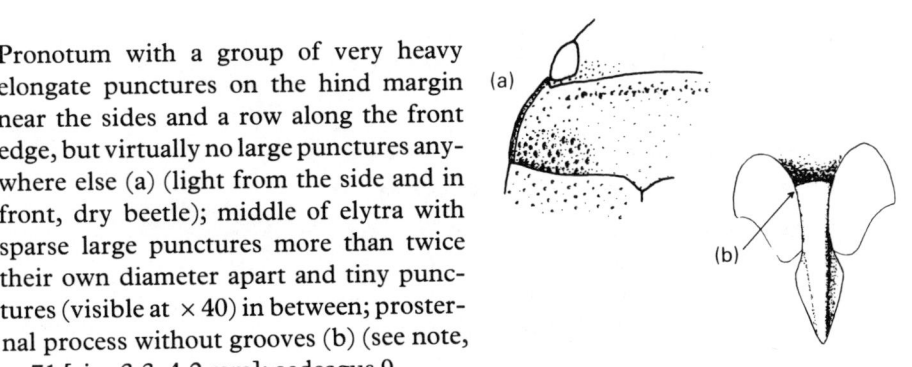

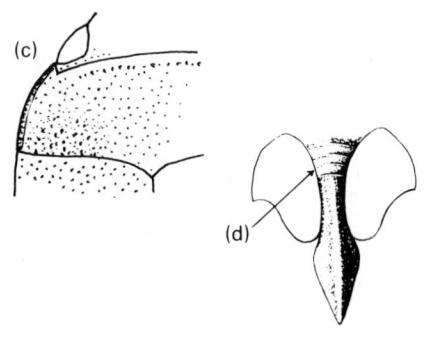

14 (12)	Sides of pronotum curving in at the rear, meeting the elytra at a distinct angle viewed directly from above (a); pronotum may be paler than elytra; [size 2.8–3.3 mm] . . . female ***H. angustatus***	
—	Pronotum broadest at the hind edge, forming a continuous curve with the sides of the elytra (b) 15	
15 (14)	Upper surface entirely dull with puncturation not obvious at × 20 . . . . . female ***H. memnonius*** (var. castaneus)	
—	Upper surface polished and shining between punctures which are obvious at × 20. 16	
16 (15)	Size 3.1 mm or less; centre of pronotum with only very fine punctures in contrast to the elytra; no longitudinal rows of punctures visible on elytra; aedeagus 11 . ***H. obscurus***	
—	Size 3.5–4.5 mm; centre of pronotum with some large punctures, only a little less marked than on elytra; elytra may or may not have lines of punctures 17	
17 (16)	Two weak longitudinal lines of large punctures on each elytron (× 12) (a); elytra flattened in profile (b); [size 3.8–4.5 mm]; aedeagus 12 . ***H. memnonius***	
—	No such lines visible on elytra; pronotum and elytra arched in profile (c); [size 3.5–4.1 mm]; aedeagus 13 . ***H. gyllenhalii*** *Note:* These two species may both be very dark; check also the aedeagi of the larger black species in the Table (p. 83).	

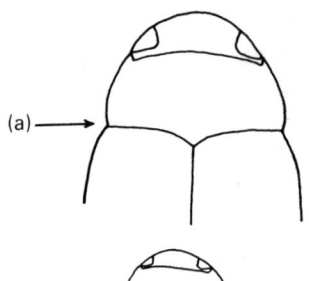

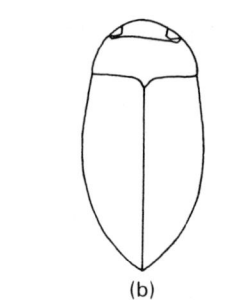

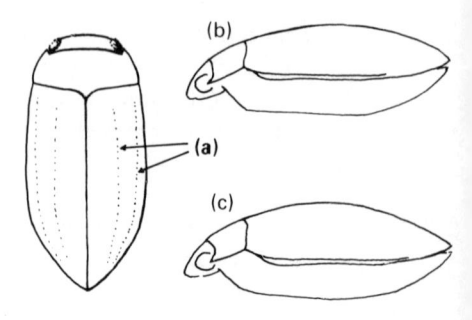

ALL BLACK, VERY DARK BROWN SPECIES

18 (9)	Tarsi and lower parts of tibiae all black viewed from above 19	
–	Legs mainly reddish or yellowish with tarsi only partly, if at all, darkened [N.B. very dark *H. memnonius* and *H. gyllenhalii* could come out here: check size—if over 3.5 mm, see couplet 17 and aedeagi 12 & 13] 20	
19 (18)	Head all black underneath; male with inner front tarsal claw toothed (a) (× 50); females as shiny as males; [size 3.3–3.5 mm]; aedeagus 14 . . . **H. morio**	
–	Head with paler markings on underside; no tooth on male front tarsal claw; aedeagus in side view with a downwardly pointing projection (b); females duller than males; [size 2.9–3.4 mm]; aedeagus 15 . . . . **H. glabriusculus**	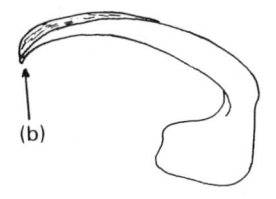
20 (19)	Elytra distinctly rounded at the sides as viewed from directly above (a); hind coxal process with sparse hairs near the midline (b) (if these hairs have rubbed off, the pits in which they originated will still be visible at × 20); pronotum entirely black . 21	
–	Front half of elytra with almost straight, parallel sides (c); hind coxal process with dense hairs around the midline (d); pronotum may have paler hind corners . 22	

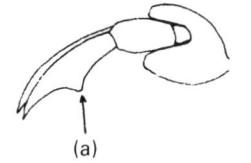

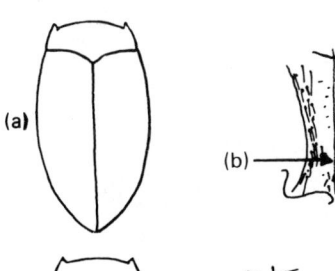

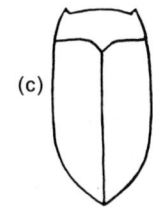

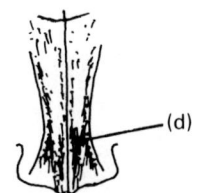

21 (20)	Elytra shining and each elytron with two faint longitudinal lines of punctures; antennal segments all pale or only slightly darkened with little contrast between segments 1–4 and 5–11; hind coxal process with a strongly wavy hind margin (a); [size 3.4–3.8 mm]; aedeagus 16. **H. longulus**	(a)
—	Elytra dull with no trace of longitudinal lines; antennal segments 5–11 with dark tips, strongly contrasted with segments 1–4 (b) (segments 5–11 are also wider than segments 1–4 in males); hind coxal process with a straight hind margin (c); [size 2.8–3.5 mm]; aedeagus 17 . . . **H. nigrita**	(b) (c)
22 (20)	Size 3.5–3.8 mm; aedeagus with long process at tip (aedeagus 18) **H. longicornis**	
—	Size 3.0–3.5 mm; aedeagus without process at tip 23	
23 (22)	*These two species can only be separated reliably using male genital characters.*	
—	Aedeagus with short end part, about 0.07 mm long (aedeagus 19); [size 3.0–3.5 mm] . . . . . . **H. melanarius**	
—	Aedeagus with longer end part, about 0.11 mm long (aedeagus 20); [size 3.2–3.5 mm] . . . . . . **H. cantabricus**	

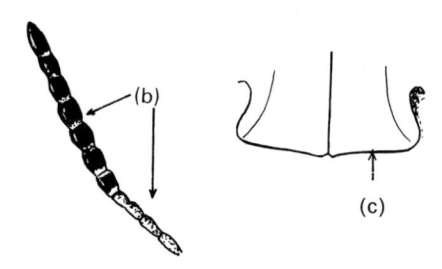

CENTRE OF PRONOTUM DARKER THAN
FRONT EDGE OF ELYTRA

24 (8)	Size 3.6–5.3 mm; elytra brown, darkening towards the tip; side keel of elytra (marked by a definite line) curving up only a little to meet hind edge of pronotum; if the beetle is laid flat on its front and viewed directly from the side (easily done if beetle is temporarily stuck on card), the depth of body below the junction of the pronotum and elytra is only about 1.5 times that above the junction (a) 25	(a)
—	Size 2.2–3.8 mm; elytra either as above or with distinctive yellow pattern; if size more than 3.0 mm, the elytra side keel is more curved towards its junction with the pronotum; viewed from the side, the depth of body below the junction of the pronotum and elytra is more than twice that above the junction (b) 26	(b)
25 (24)	Size 3.6–4.5 mm; antennal segments 3 & 4 each much shorter than segment 5 and each less than twice as long as broad (a) (× 40); elytra densely hairy (these may rub off); females may be duller than males; aedeagus 21 . . **H. erythrocephalus**	(a)
—	Size 4.0–5.3 mm; antennal segments 3 & 4 only a little shorter than segment 5 and each more than twice as long as broad (b); elytra with sparse hair; females never markedly duller than males; aedeagus uniquely expanded; aedeagus 22 . . . **H. rufifrons**	(b)

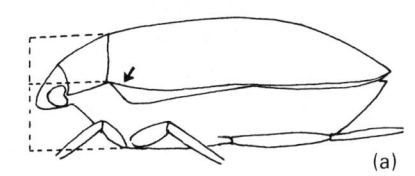

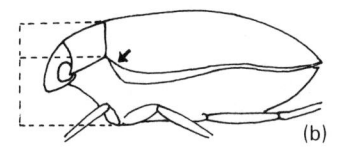

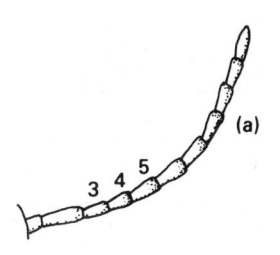

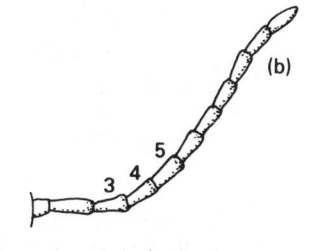

26 (24)	Pronotum entirely black or very dark brown; elytra brown, darkening towards apex but not with a markedly paler band at the shoulder (a) 27	
—	Pronotum black or brown, often with a broad paler side margin; elytra with a paler shoulder mark (b) or band, or strong pale pattern 31	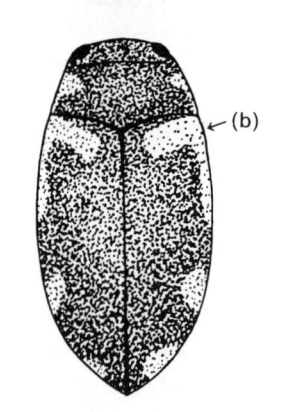
27 (26)	Size 2.2–2.6 mm; hind edge of hind coxal process wavy (a) (light from the side, view on a dry beetle); male with a lobe on the inner front tarsal claw (b) ($\times 50$); aedeagus 23 . . . . . **H. neglectus**	
—	Size 2.5–3.8 mm; hind edge of hind coxal process straight (c); no lobe on front tarsal claw of male 28	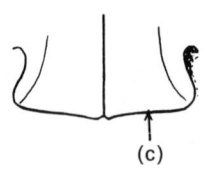
28 (27)	Legs black; aedeagus in side view with a downwardly pointing projection (see cpt 19) . . . . . . **H. glabriusculus**	
—	Legs reddish with tarsi only partly, if at all darkened; aedeagus not as above . . 29	

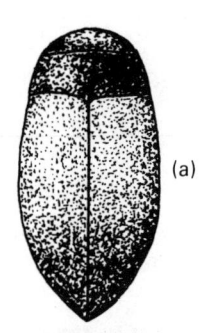

29 (28)	Size 3.0–3.8 mm; sides of pronotum strongly curved, especially at rear, and the hind edge of the pronotum slightly wider than the front edge of the elytra (a); aedeagus broad; aedeagus 24 . . . . . **H. elongatulus**	
—	Size 2.5–3.3 mm; sides of pronotum less curved, hind edge of pronotum not wider than front edge of elytra (b); aedeagus narrow 30	
30 (29)	Size 2.8–3.3 mm; aedeagus narrowly rounded at tip; aedeagus 25 . **H. tristis**	
—	Size 2.5–2.9 mm; aedeagus broadly rounded at tip; aedeagus 26 . . . . . **H. umbrosus**	
31 (2) (26)	Neck of prosternal process with a distinct step but no grooves (a) (see note p. 71); elytra black and dull, usually with distinctive yellow pattern (b) which may be reduced to the sides only; [size 3.3–3.8 mm]; male with one front tarsal claw much thicker than the other (×50) (c); aedeagus 27 . . . . . **H. palustris**	
—	Neck of prosternal process with grooves (d); elytra black and dull with only faint shoulder marks or shining brown with yellow pattern; male front tarsal claws alike 32	

32   Size 3.3–3.8 mm; elytra brown and shining
(31)   with yellow pattern (a); last abdominal
       segment (underside) densely punctured
       (about two reticulation meshes apart) but
       shining in between; aedeagus 28.   . . .
                                          H. incognitus

—   Size 3.0–3.4 mm; elytra black and dull
       with only faint pale shoulder and side
       markings (b); last abdominal segment
       sparsely punctured (about four reticula-
       tion meshes apart) and appearing dull;
       aedeagus 29 .   .   .   .   .   . **H. striola**

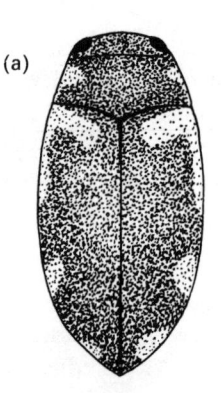

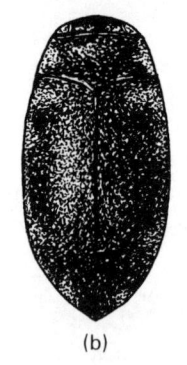

This key is based on Foster & Angus, 1985, by the authors' kind permission.

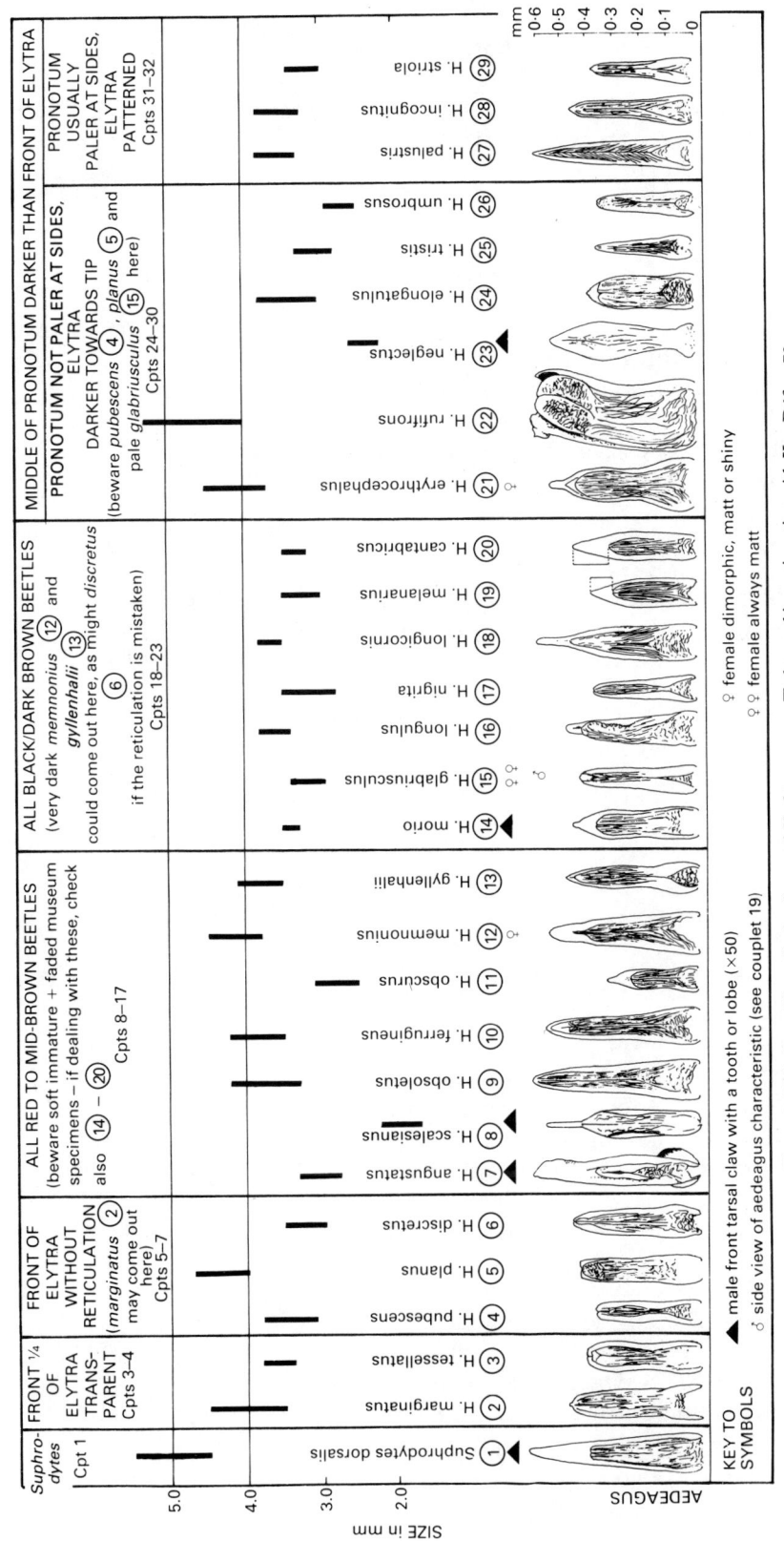

Figure 9. Tabular guide to *Suphrodytes* and *Hydroporus* spp. To be used in conjunction with Key D4f, p. 70

Key D4g *Potamonectes*

Genus Characters: no furrows in pronotum; linear coloration (may coalesce); no large punctures on underside of thorax and abdomen.
Size range 4.0–5.2 mm.
4 spp.

1   Elytra each with a small tooth (a) projecting from the margin near the tip (if no teeth visible from above, turn the beetle over and check again by looking from below) 2

—   No tooth near tip of the elytra; elytra dark with 7–8 yellowish lines (b); [size c. 4.5 mm] . . . . **P. griseostriatus**

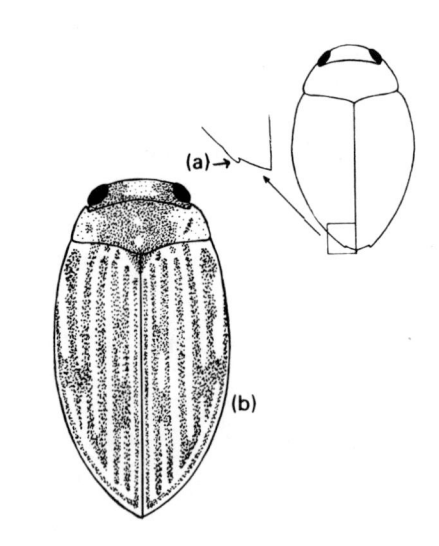

2   Size 4.0–4.3 mm; blotches on the hind
(1) margin of the pronotum extending forwards half-way across the pronotum (a); sides of pronotum curving in only a little at rear (b), to meet elytra at slight angle .
     P. assimilis

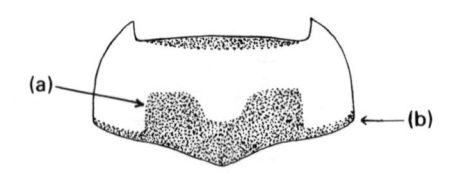

—   Size 4.5–5.2 mm; blotches on the hind margin of the pronotum not extending more than 1/3 way across the pronotum (c); sides of pronotum very strongly curving in at rear (d), meeting elytra at a distinct angle 3

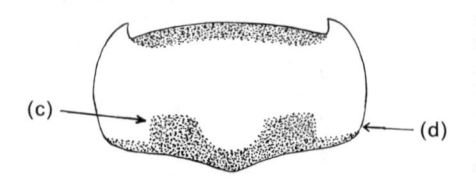

3   **Note:** The *depressus-elegans* complex. Intermediate
(2)   forms are found where *depressus* and *elegans* occur
together; examination of the male genitalia and claws
is needed for certain identification.

—   Elytra usually predominantly dark, with yellow lines (a) but can be light; males with the claws of the front tarsi hooked (b), and tip of the aedeagus (the middle part of the genitalia) wide (c) . . . .
           P. depressus depressus

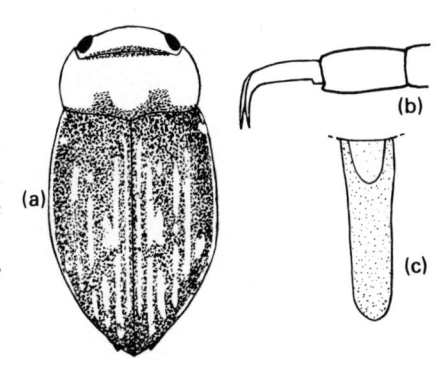

—   Elytra usually light, with twelve yellow spots and yellow lines (d) but can be very dark; males with the front tarsal claws evenly curved (e), and tip of the aedeagus narrow (f) . .   ***P. depressus elegans***

Note: if the longitudinal lines are obscured, check again for the teeth at the end of the elytra; if no teeth present, check *Stictotarsus* (Key D4, p. 56).

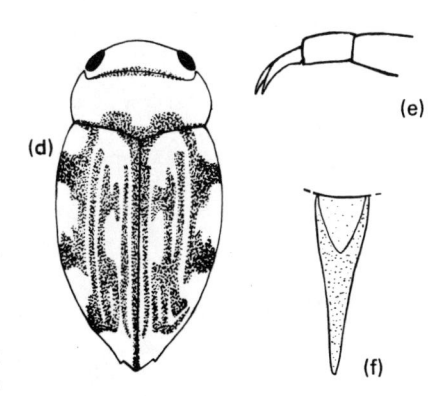

Key E   CHRYSOMELIDAE

Family Characters: antennae inserted close together on front of head; all tarsi 4-segmented. 3 subfamilies, associated with water plants, willows, alders, etc. key out here.
See Joy (1932) for keys to species.

1	Head narrow behind the eyes, forming a definite "neck" (a); pronotum narrowing behind and much narrower than the front of the elytra (b) . .Donaciinae . .	2
—	Head without a definite "neck"; pronotum only a little narrower than the elytra, or if much narrower, then not narrowing behind (c)	5
2 (1)	Antennae almost as long as the beetle (a)	3
—	Antennae not so elongate . . . . . . **terrestrial Donaciinae**	
3 (2)	Elytra with spines near the tip (a); 3rd segment of the tarsi minute, the 4th elongate (b) [size 4.5–8.5 mm] ***Macroplea*** 2 spp.	
—	Elytra without spines; 3rd segment of the tarsi lobed (c)	4

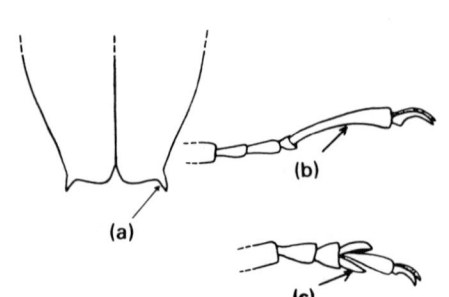

4 (3)	Tip of the elytra rounded (a); ridges along the midline of the elytra diverging (leaving the edges of elytra) near the tip (b) (view from behind); legs short and broad, when stretched out behind, the hind tibia only just over-reaching the tip of the elytra [size 5.0–12.0 mm] **Plateumaris** 4 spp.	
—	Tip of the elytra cut off squarely (c); ridges on the midline of the elytra staying on the edge of the elytra right to the tip (d); legs long and slender (in most species, the hind tibia far over-reaching the tip of the elytra) [size 6.0–13.0 mm] . . **Donacia** 15 spp.	
5 (1)	Hind femur much broader than the middle femur . . . . . .**Halticinae** (flea beetles)	
—	Hind femur not much broader than the middle femur . . . . . **Galerucinae**	

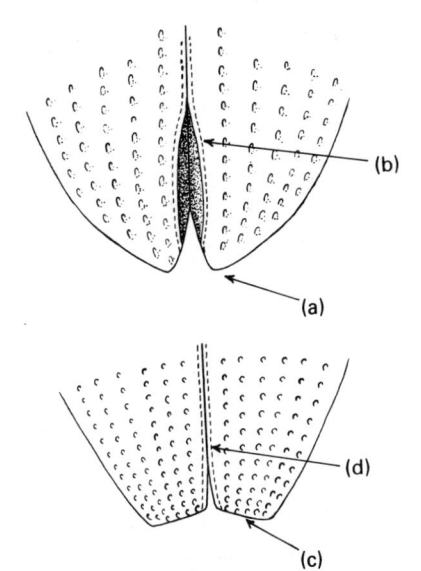

Key F   **ELMIDAE (ELMINTHIDAE)**

Family Characters: legs and claws long, 5th segment of all tarsi much longer than rest and bulbous at tip; pronotum and/or elytra with tubercules or sharply-defined longitudinal ridges.
Size range 1.3–4.75 mm.
8 genera.

Note: Sizes are from front of the pronotum to tip of the elytra, as the head is often hidden. If beetle is covered all over with dense hairs, turn to Key H, p. 141.

1   All legs very long, at least as long as the entire beetle; antennae very short, apparently clubbed; two shiny, hairy bumps (tubercles) on the pronotum and two on the shoulders of the elytra (a) [size 2.9–3.2 mm] . . . .
Macronychus quadrituberculatus

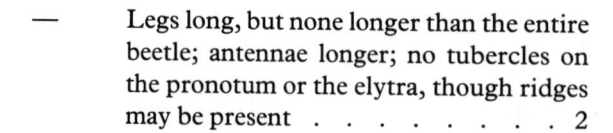

—   Legs long, but none longer than the entire beetle; antennae longer; no tubercles on the pronotum or the elytra, though ridges may be present 2

2   Size 3.75–4.75 mm; pronotum with a shallow longitudinal furrow in the mid-line but no strong ridges; elytra with strong ridges, two reaching only half-way (a). .
(1)   *Stenelmis canaliculata*

—   Size 3.2 mm or less; sculpture different from above 3

3   Pronotum with 2 sharply-defined ridges
(2)   (see couplets 4–6 for Figs.) 4

—   Pronotum with a definite lateral border, but no ridges (a) *Riolus* and *Normandia*
                           Key F1, p. 90

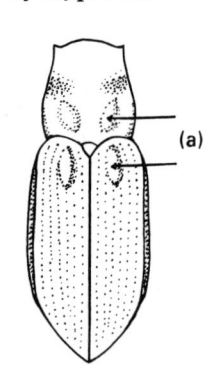

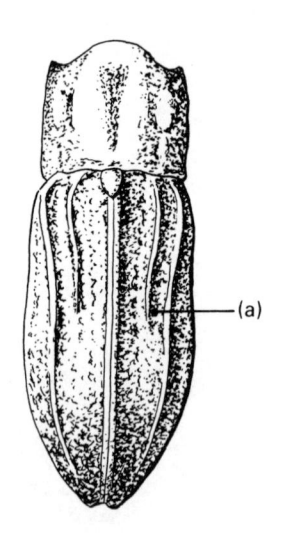

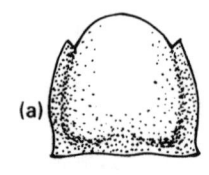

4 (3)	Ridges on the pronotum curving in behind to virtually meet (a); 2 marked ridges on each elytron [size 1.9–2.2 mm] . . . . ***Elmis aenea***	
—	Ridges on the pronotum remaining well separated throughout their length . . 5	
5 (4)	Elytra with longitudinal furrows formed of rows of punctures, but no strong ridges (a); [size 2.9–3.2 mm] . . . . ***Limnius volckmari***	
—	Elytra with strong longitudinal ridges in addition to rows of punctures; size less than 2.1 mm 6	
6 (5)	Elytra each with one ridge, beginning half-way between the ridge on the pronotum and the lateral margin (a) [size 1.3–1.5 mm] ***Esolus parallelepipedus***	
—	Elytra each with 3 ridges, the inner one being a continuation of the ridge on the pronotum (b) . . . . . ***Oulimnius*** Key F2, p. 91	

Key F1   *Riolus* and *Normandia*

Genera Characters: no ridges on the pronotum; 1 or 3 ridges on each elytron.
Size range 1.75–2.30 mm.
Riolus 2 spp.
Normandia 1 sp.

1   Antennae and legs mid-brown to black; body elongate; each elytron with three ridges (a), which may be conspicuously hairy; [size 1.85–2.30 mm] . . . . .
    Riolus subviolaceus

—   Antennae yellow or red; legs brown; body short and broad; elytra with one or three ridges; [size 1.75–1.95 mm] 2

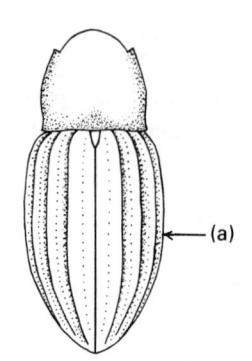

2   Three equally raised ridges on each
(1) elytron (a) . . . . . ***Riolus cupreus***

—   One conspicuously raised ridge on each elytron (b), starting level with the lateral margin of the pronotum . . . . . .
    Normandia nitens

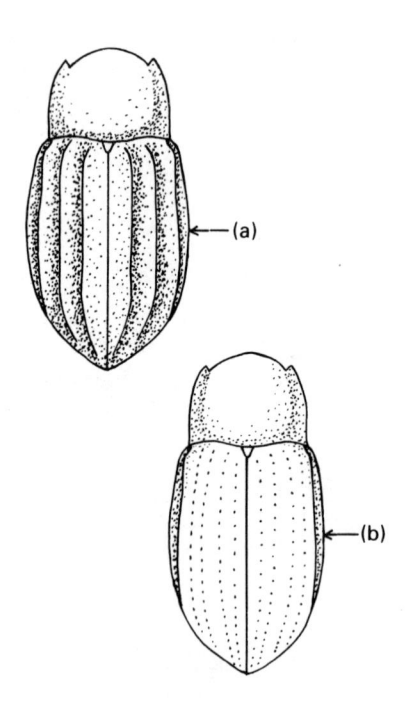

Key F2   *Oulimnius*

Genus Characters: pronotum with 2 ridges and each elytron with 3; all ridges with "beaded" appearance.
4 spp.

Note: Examination of the genitalia (see note at the front of the key, p. 5) is needed for certain identification, this is most effectively done on a slide under a coverslip at very high magnification (×80). Genitalia are redrawn here from Berthélemy (1979) in ventral view and Parry (1980).

1   Elongate body shape (a); length more than 2 × body width; genitalia as 1, 3, 4 below
. . . . 2

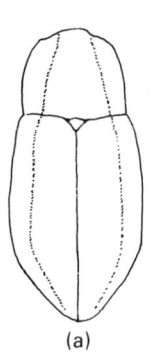

(a)

—   Body relatively broad (b); length less than 2× body width; genitalia as 2 below, female lobes very short with a strong spine on inside at tip; [size 1.5–1.7 mm] . . .
O. troglodytes

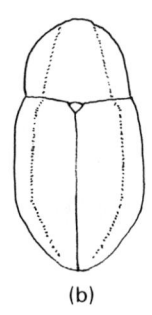

(b)

2   **Rely on identification of males only**

—   Male with asymmetrical aedeagus (genitalia 4 below); elytra dark, brassy; [size 1.4–1.9 mm] . . . . . ***O. rivularis***

—   Aedeagus symmetrical in ventral or dorsal view (beware side view of *O. major* 3— check width of parameres); elytra reddish
. . . . 3

3   Aedeagus with tooth on dorsal edge (3), seen in side view, and wide parameres; [size 1.8–2.0 mm] . . . . . ***O. major***

—   Aedeagus without dorsal tooth; narrow parameres (1); [size 1.7–1.9 mm] . .
O. tuberculatus

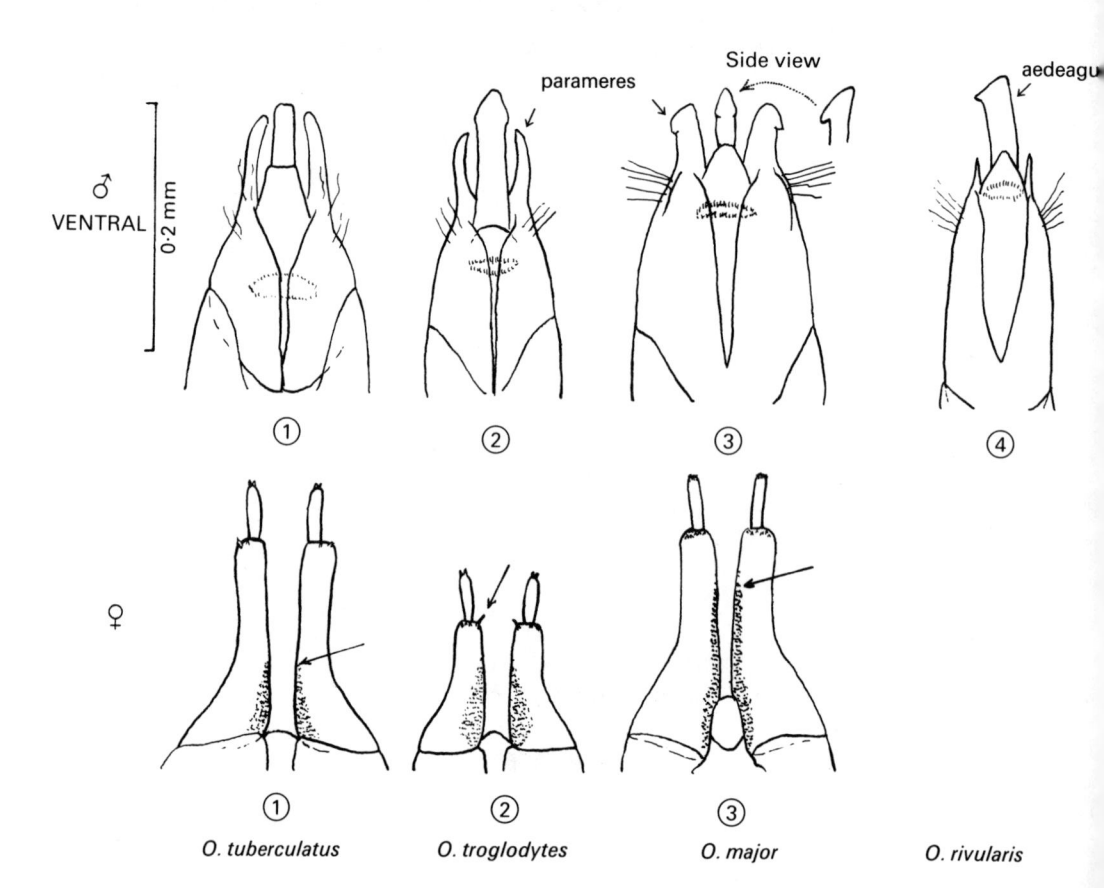

① *O. tuberculatus*   ② *O. troglodytes*   ③ *O. major*   *O. rivularis*

KEY G
HYDROPHILIDAE AND HYDRAENIDAE

Family Characters: palps from 2/3 to 4 times as long as the antennae, which are clubbed.
  Hydrophilidae: size range 1.0–48.0 mm.
   6 subfamilies containing 21 genera
  Hydraenidae: size range 1.0–3.0 mm.
   3 genera

1   Size 38.0–48.0 mm; colour silvery black; underside with mid-line keel extending beyond the hind coxae to end in a long point (a); Britain's largest water beetle .
   Hydrophilus piceus
   (HYDROPHILIDAE, Hydrophilinae)

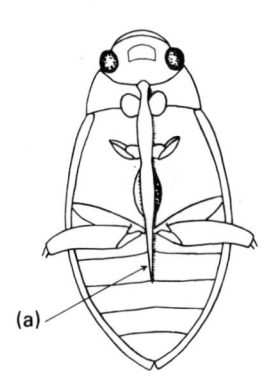

—   Size less than 20.0 mm 2

2   15.0–18.0 mm; black shining beetle with
(1) faint longitudinal lines on the elytra; mid-line keel (underside) extending from the head to the hind coxae (a) . . . . . .
   Hydrochara caraboides
   (HYDROPHILIDAE, Hydrophilinae)

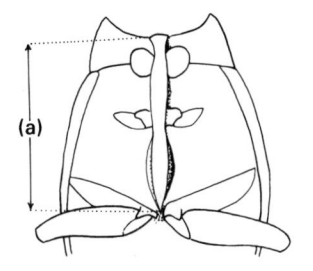

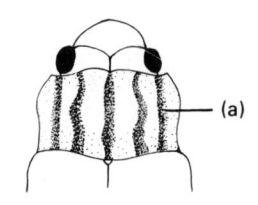

—   Size 10.0 mm or less 3

3   Pronotum with five furrows running full
(2) length (a); [size 1.9–7.5 mm] . . . . .
   Helophorus
   (HYDROPHILIDAE, Helophorinae)
   Key G1, p. 101

—   Pronotum without five such grooves, but pits may be present 4

4	Pronotum narrowing towards rear (a);
(3)	pronotum with large pits, depressions or grooves (b) 5

— Pronotum not markedly narrower behind, and without pits (though fine punctures may be present) 8

5	Size 6.0–7.0 mm; front of the head with a
(4)	deep, wide indentation (a); pronotum with shallow depressions; beetle yellowish with brown blotches (b) . . . .

Spercheus emarginatus
(HYDROPHILIDAE, Sperchinae, 1 sp.)

— Size 1.0–4.5 mm; front of head smooth, or with a narrow V-shaped notch, or a very shallow indentation 6

6	Underside with conspicuous round punc-
(5)	tures (a); pronotum with characteristic depressions (b); [size 2.2–4.5 mm] . . .

Hydrochus
(HYDROPHILIDAE, Hydrochinae)
Key G2, p. 111

— Underside without round punctures (though other sculpture may occur); pronotum with depressions in a different arrangement (see next couplet); [size 1.0–2.5 mm] 7

7 Palps much longer than antennae (see note
(6) p. 5), as long as the head and pronotum
(a); pronotum with a longitudinal shallow
depression on each side (b) . **Hydraena**
(HYDRAENIDAE) Key G10, p. 131

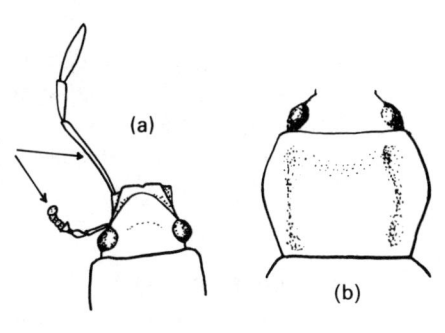

— Palps about equal to the antennae (c); pronotum with two transverse shallow pits (d), or a midline longitudinal groove (e) .
Ochthebius
(HYDRAENIDAE) Key G11, p. 135

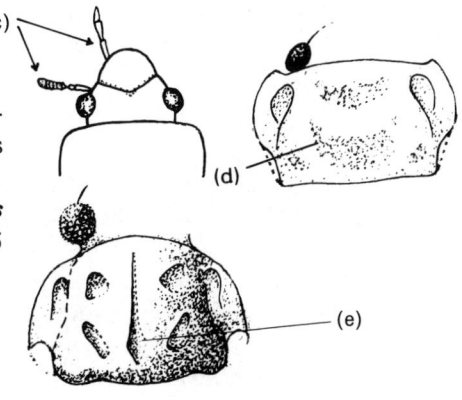

8 Head, antennae and palps hidden from
(4) above; pronotum crinkled on the front
edge (a); [size 1.5–1.8 mm] . . . . .
Georissus crenulatus (terrestrial)
(HYDROPHILIDAE, Georissinae, 1 sp.)

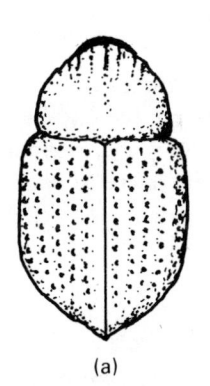

— Head visible from above; pronotum smooth 9

9	Larger punctures on elytra arranged in 10–20 more or less regular rows; in *some* genera, puncture rows are marked by about 10 longitudinal grooves down each elytron (a) 10
(8)	

—	Punctures on the elytra apparently arranged at random with, at most, 2–3 indistinct lines of larger punctures among general puncturation; no longitudinal grooves down elytra except perhaps for one either side of midline (b) 16
	Note: If the palps are very long (1.4 to 2× longer than the antennae), follow the second lead.

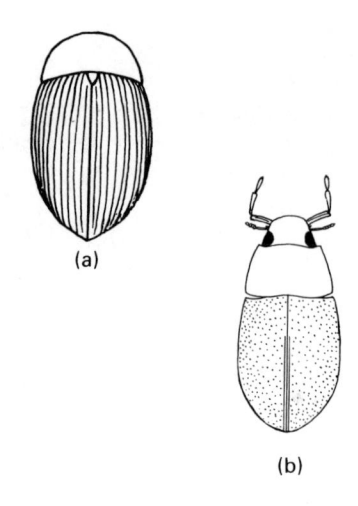

10	Middle and hind legs with long pale swimming hairs on both the tibiae and the tarsi (a); underside with a large pit in the midline just in front of the hind coxae (b); elytra with 10 longitudinal rows of pits and grooves, plus an additional short row on the shoulders between the 1st and 2nd rows; elytra yellowish; general appearance characteristic (c); [size 3.5–5.5 mm] . .
(9)	

Berosus
(HYDROPHILIDAE, Hydrophilinae)
Key G4, p. 117

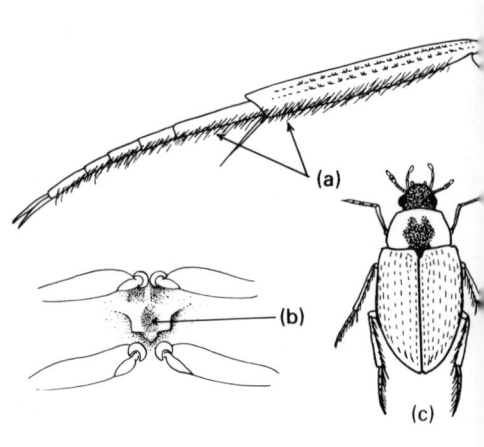

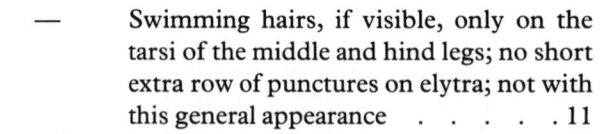

—	Swimming hairs, if visible, only on the tarsi of the middle and hind legs; no short extra row of punctures on elytra; not with this general appearance 11

11	Size 1.3–4.4 mm; upper surface uniform colour or with elytra completely or partly paler than pronotum and head . . . 12
(10)	

—	Size 5.0–10.0 mm; upper surface uniform black or dark brown, shining 15

12 (11)	Head, viewed from above, with labrum clearly visible, its width (front to back) greater than the width of the last segment of the palps (a); pronotum with irregular dark spot in the centre (b); elytra paler than centre of pronotum with puncture rows usually marked by lines of dark pigment; [size 2.0–4.5 mm] . . *Laccobius* (HYDROPHILIDAE, Hydrophilinae) Key G7, p. 123	
—	Labrum not visible, or only narrowly visible, from above (c); colours not as above, although the sides of the pronotum and part or all of the elytra may be pale 13	
13 (12)	Front tibia with deep indentation near the joint with the tarsus (a); elytra shining between the punctures; [size 1.5–2.0 mm] ***Megasternum obscurum*** (HYDROPHILIDAE, Sphaeridiinae), (terrestrial)	
—	Front tibia club-shaped (b) or cut across obliquely at the end (c); elytra dull or shining 14	
14 (13)	Upper surface covered with fine sparse hairs; elytra heavily punctured between rows; [size 1.6–2.0 mm] . . . . . . ***Cryptopleurum*** 3 spp. (HYDROPHILIDAE, Sphaeridiinae) (terrestrial, see Joy, p. 289)	
—	Upper surface without hairs; elytra only finely punctured, or shining between rows; [size 1.3–4.4 mm] . . ***Cercyon*** (HYDROPHILIDAE, Sphaeridiinae) Key G3, p. 113	

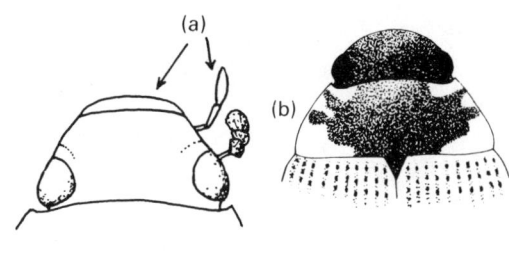

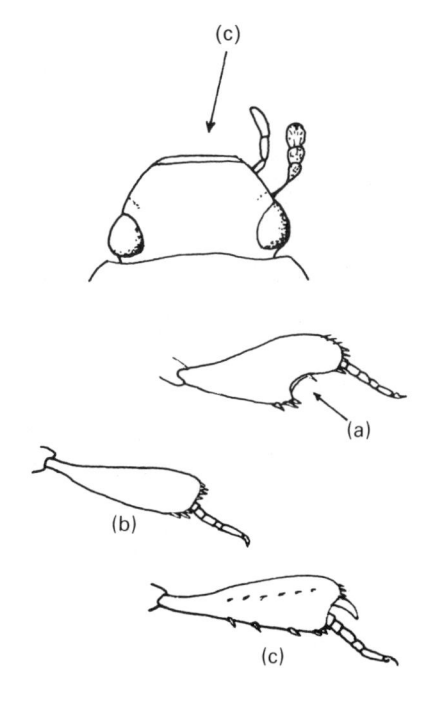

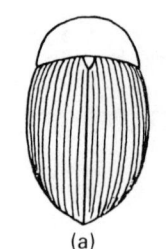

15 (11)	Size 5.0–8.0 mm; elytra each with about 10 distinct longitudinal grooves (a). . . ***Hydrobius fuscipes*** (HYDROPHILIDAE, Hydrophilinae)	
—	Size 8.0–10.0 mm; grooves on elytra not apparent, except occasionally at the hind tip . . . . . ***Limnoxenus niger*** (HYDROPHILIDAE, Hydrophilinae)	
16 (9)	Elytra truncate so that some abdominal segments are visible from above (a) (beware swollen specimens preserved in alcohol, look for cut-off end of elytra); no groove on either side of midline on elytra; [size 1.0–2.5 mm] . . . ***Limnebius*** (HYDRAENIDAE) Key G9, p. 129	
—	Elytra not truncate, completely covering abdomen; if beetle less than 2.5 mm, elytra with a groove on either side of the midline in the rear half (light from rear) (b) . . 17	
17 (16)	Minute, domed black beetle; width across the elytra equal to their length (a); first visible abdominal segment (underside) with a fringe of long, pale hairs extending back over the next segment (b) [size 1.0–1.5 mm] ***Chaetarthria seminulum*** (HYDROPHILIDAE, Hydrophilinae)	
—	Beetle larger, 2.0–7.5 mm, and more elongate (c); no pale fringe of hairs on the first abdominal segment 18	

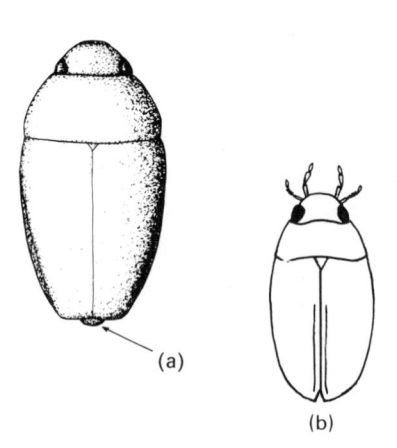

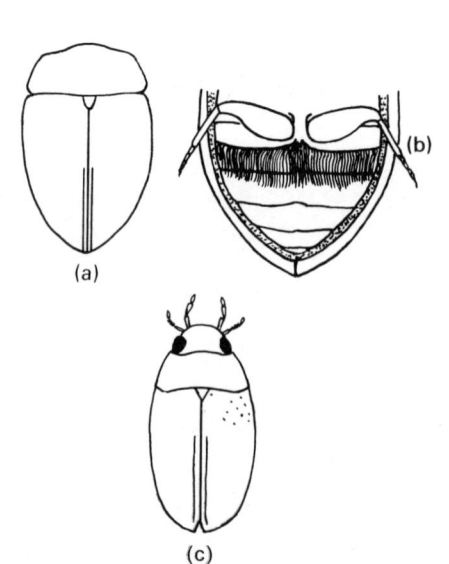

18	Palps about equal in length to or slightly shorter than antennae (a) (see note p. 5) 19
(17)	

— Palps much longer than antennae, from 1.4× to more than twice as long (b) . 22

19	Head, viewed from above, with the labrum clearly visible, its width (front to back) greater than the width of the last segment of the palps (a); each elytron with punctures in about 20 irregular rows; [size 2.5–4.0 mm] . . . . . . *Laccobius* (HYDROPHILIDAE, Hydrophilinae) Key G7, p. 123
(18)	

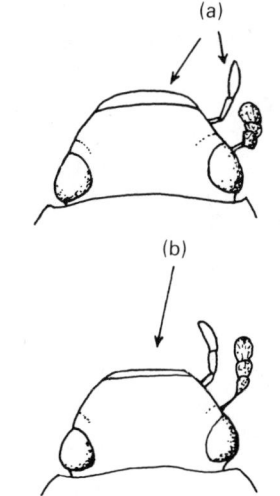

— Labrum not visible from above, or only very thinly visible (b); puncturation of elytra not in 20 rows, but there is a groove on either side of the midline 20

20	Size 2.0–3.0 mm . . . . *Anacaena* and *Paracymus* (HYDROPHILIDAE, Hydrophilinae) Key G8, p. 127
(19)	

— Size 4.0–7.5 mm 21

21	Mesoscutellum (at junction of the elytra and the pronotum) much longer than broad (a); spines on the tibiae long and stout (b); elytra almost always with conspicuous pale spots (c); [size 4.0–7.5 mm] ***Sphaeridium*** 3 spp. (HYDROPHILIDAE, Sphaeridiinae) (terrestrial, Joy p. 288)
(20)	

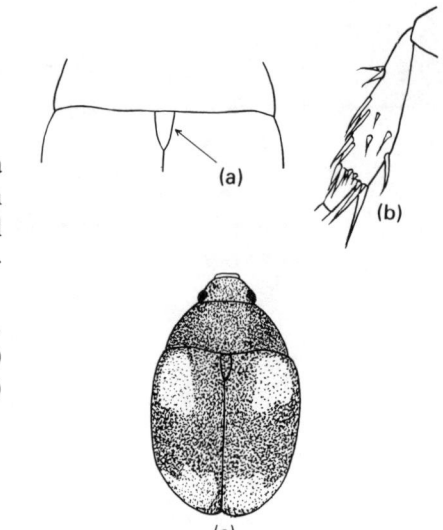

— Mesoscutellum a little longer than broad (d); spines on the tibiae short and fine in comparison with the tibial spurs (e) elytra uniformly dark; [size 4.5–5.0 mm] . . . ***Coelostoma orbiculare*** (HYDROPHILIDAE, Sphaeridiinae)

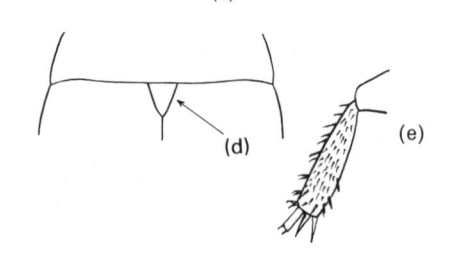

22	Elytra with a distinct groove on either side of the midline in the rear half (a); palps less than twice as long as antennae . . . 23
(18)	

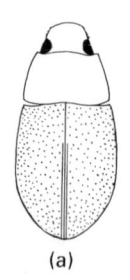

— Elytra without a groove on either side of midline; palps about twice as long as antennae; [size 4.5–6.5 mm] ***Helochares*** (HYDROPHILIDAE, Hydrophilinae) Key G5, p. 119

A Key to the Adults of British Water Beetles

23	First visible segment of the palps bending out, more curved than the next segment (a); all tarsi with 5 segments (not counting claws), the short first segment visible at ×12 or more (b) elytra yellow to dark brown; [size 3.0–7.0 mm] . . ***Enochrus*** (HYDROPHILIDAE, Hydrophilinae) Key G6, p. 120
—	First visible segment of the palps almost straight, less curved than the next segment (c); all tarsi with 4 segments (d); mature beetle is all black with chestnut margins on the elytra; (soft immature specimens may be brown) [size 4.0–4.5 mm] . . . ***Cymbiodyta marginella*** (HYDROPHILIDAE, Hydrophilinae)

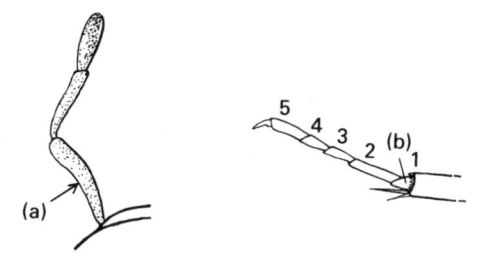

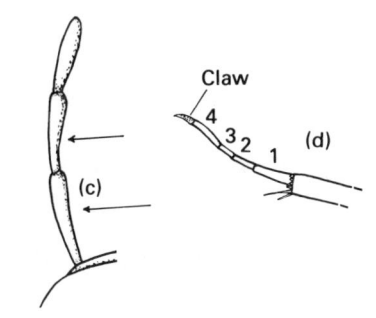

Key G1 *Helophorus*

Genus Characters: pronotum with longitudinal ridges and furrows.
  Side range 1.9–7.0 mm.
  17 aquatic spp. key out here.

Note: Certain identification of several of the species is possible only by examination of the male genitalia. *Helophorus* cannot be sexed by external characters so all specimens will need to be examined. See note on extraction and display of genitalia p. 5. Live specimens will extrude the genitalia if the abdomen is gently squeezed.

1	Beetle all black, with shiny bumps (tubercules) on the elytra (a) [size c. 3.0 mm] . **H. tuberculatus**
—	Beetle in part brown or yellowish and without tubercules on the elytra. . . 2

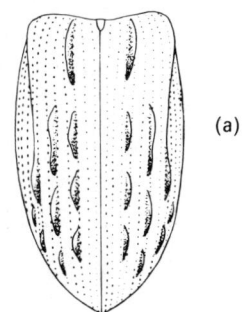

2 (1)	A short row of punctures (a) present on the front of the elytra between the 1st and 2nd longitudinal rows from the middle . .	3
—	All rows of punctures on the elytra running almost full length	6
3 (2)	Spaces between alternate rows of punctures on the elytra developed into obvious ridges (a) surmounted by rows of stiff hairs; tarsi with short, stiff spines on the top and sides of the segments (b) . . . **Terrestrial species** *H. nubilus* *H. rufipes* *H. porculus*	
—	Spaces between alternate rows of punctures on the elytra slightly convex (c) with a few fine hairs; tarsi with only fine swimming hairs on top and sides (difficult to see) (d)	4

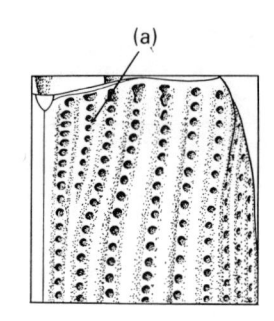

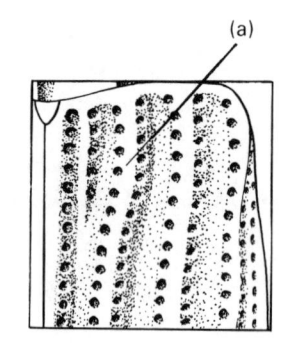

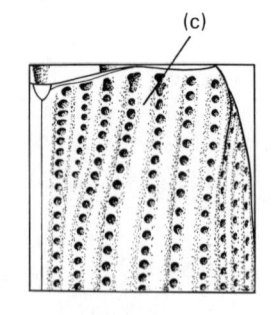

A Key to the Adults of British Water Beetles

4
(3)   Last segment of the maxillary palps symmetrical when viewed from any angle, inner and outer faces equally curved (a); sides of elytra seen from below (elytral flanks) about as wide as the epipleurs at the level of the hind coxa—see box below; last abdominal segment smoothly rounded (the hind margin may be partly obscured by hairs) male genitalia (a) p. 104 [size 4.0–5.0 mm] . . . ***H. alternans***

—   Last segment of maxillary palps asymmetrical when viewed from some angles, with inner face much straighter than outer face (b); elytral flanks not visible from below or much narrower than the epipleurs at the level of the hind coxa; hind margin of the last abdominal segment toothed (c). 5

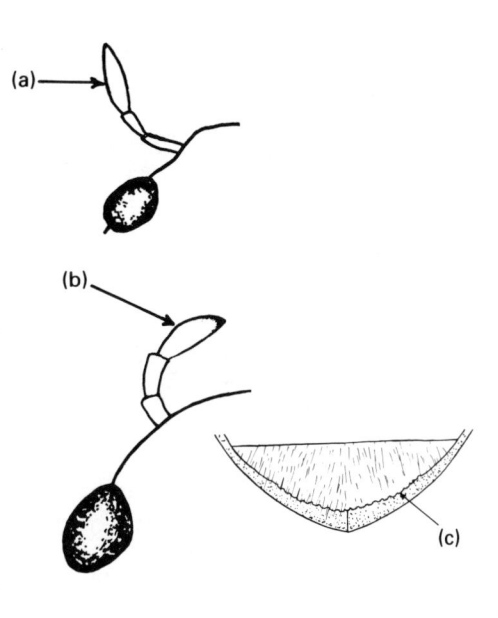

ELYTRAL FLANKS

Elytral flanks (ef) are defined here as those parts of the **upper surface** of the side margins of the elytra which can be seen when the beetle is viewed **from below**.

Look at the level of the hind coxae

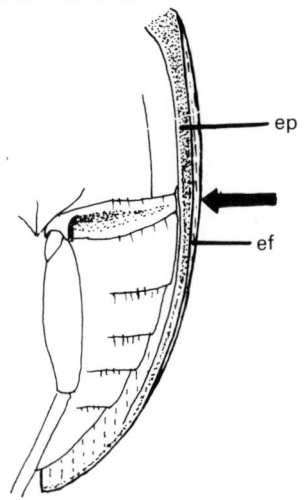

ELYTRAL FLANKS (ef)
"BROADLY VISIBLE"
Equal in width to
the epipleurs (ep)

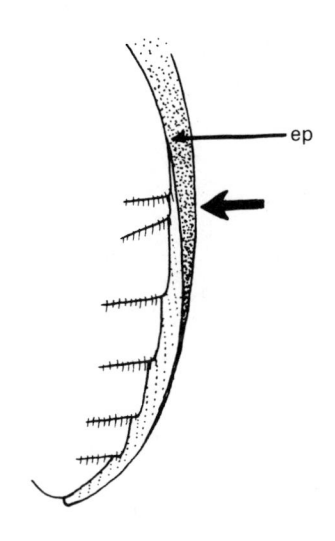

ELYTRAL FLANKS
"NOT VISIBLE"

5 (4)	"Teeth" of the last abdominal segment small, as shown (a); male genital capsule 0.7–0.9 mm long, parameres not curving out at tip (genitalia (b) below); [size 4.5–6.5 mm] . . . . . **H. aequalis**	(a)
—	"Teeth" of the last abdominal segment deeper, as shown (b); male genital capsule 1.1–1.2 mm long, parameres slender, curving out at tip (genitalia (c) below); [size 5.5–7.0 mm] . . . **H. grandis**	(b)

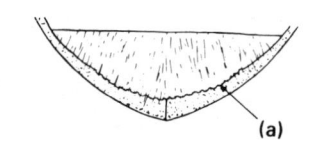

Note: hairs may obscure the view of the abdominal teeth from below; try viewing from above by moving the elytra apart at the tip and placing the beetle against a white background (× 50 or more).

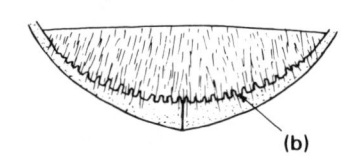

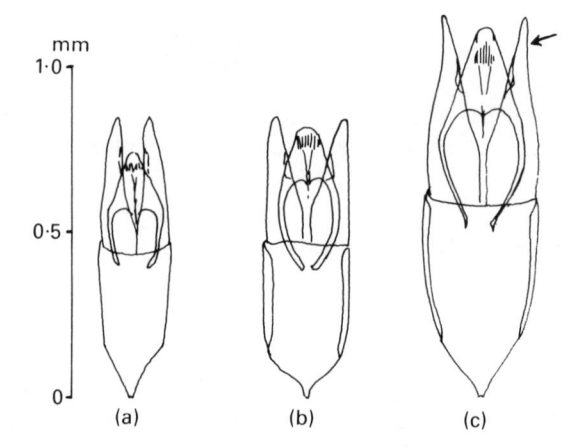

Helophorus male genital capsules: aquatic species with an extra row of punctures on the elytra.
(a) *H. alternans*
(b) *H. aequalis*
(c) *H. grandis*

A Key to the Adults of British Water Beetles

6   **SPECIES WITHOUT AN EXTRA PUNCTURE ROW ON ELYTRA**
(2)

Note: from this point on, use Fig. 10, p. 110 at the end of the key to make comparisons between the male genitalia etc. of similar species.

Last segment of maxillary palp approximately symmetrical viewed from any side, inner and outer faces having similar curvature (a); sides of elytra seen from below (=elytral flanks) very broad, as wide as or wider than the epipleurs at the level of the hind coxae (see box at cpt 4) (b) . . . 7

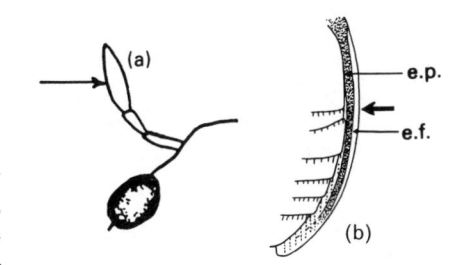

— Last segment of maxillary palps asymmetrical, the inner face much straighter than the outer (c); elytral flanks either as broad as the epipleurs, or much narrower, or not visible at all from below . . . 8

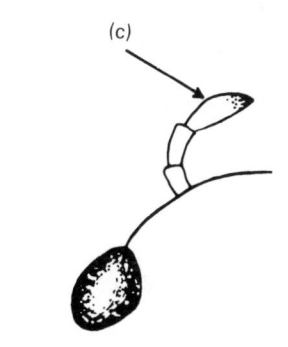

7   Pronotum with strongly wavy side margins, curving in then out towards hind margin (a); last segment of maxillary palp short, its length only about twice its width at the widest point (b); male genitalia (1); [size 2.5–3.5 mm] . . **H. arvernicus**

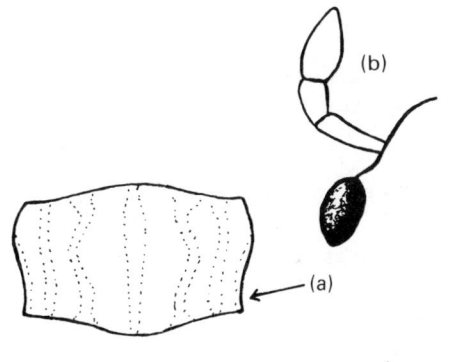

— Pronotum sides not wavy, curving in evenly towards hind margin (c); last segment of maxillary palp long, $2\frac{1}{2}$–3 times longer than its maximum width (d); male genitalia (2); [size 1.9–3.5 mm] . . . .
H. brevipalpis
(compare also genitalia (3) (4) (6) & (7))

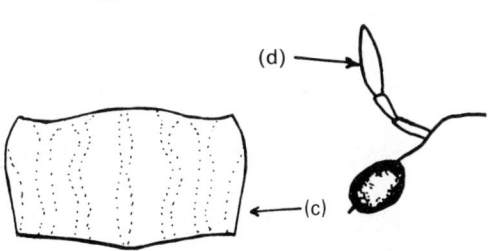

8	The Y-shaped groove on the head with a narrow, parallel-sided stem, the front end narrower than the width of the last segment of the palps (a) 9
(6)	

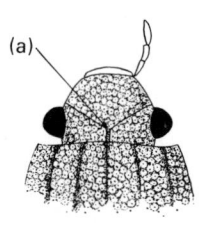

— Y-shaped groove with the stem widening at the front to be as wide as the width of the last segment of the palps (b) 11

Note: if difficulty arises here, use the table to check elytral flank, size and male characters between the two groups of species.

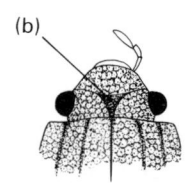

9	Head and most of the pronotum appearing granulate; head without a short groove on either side of the stem of the Y-shaped groove (a) 10
(8)	

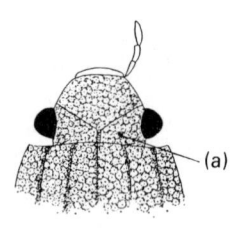

— Head and pronotum virtually without granulation, appearing brightly polished; head with an extra short groove on either side of the stem of the Y-shaped groove (b); male genitalia (3); [size 2.5–3.5 mm] .
H. nanus

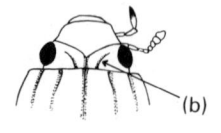

10	Elytral flanks (ef) as broad as the epipleurs (ep) at the level of the hind coxae (a); male genitalia (4) with outer edges of the parameres (lobes) curved and the struts of the aedeagus (middle part) longer than its tube; [size 3.0–4.5 mm] **H. strigifrons** (beware *fulgidicollis* and *dorsalis* here, see genitalia (6) and (7))
(9)	

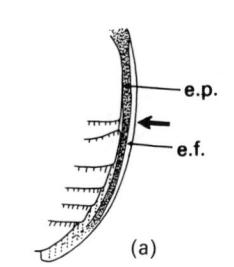

—	Elytral flanks (ef) not visible from below (b); male genitalia (5) with outer edges of the parameres straight and the struts of the aedeagus about the same length as the tube; [size 3.0–4.5 mm] . **H. laticollis** (compare genitalia (8)–(13))

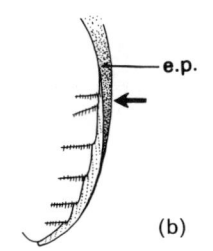

11	Elytral flanks viewed from below at least half as wide as the epipleurs at the level of the hind coxae 12
(8)	

—	Elytral flanks hardly, if at all, visible from below 13

| 12 | Elytra dark brown with conspicuous pale |
|(11)| spots near the tip and near the shoulder (a); male genitalia (6); [size 3.0–4.0 mm] . **H. dorsalis** |

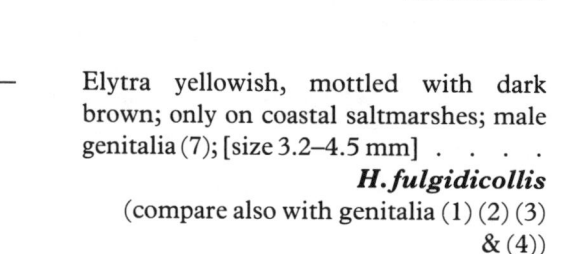

— Elytra yellowish, mottled with dark brown; only on coastal saltmarshes; male genitalia (7); [size 3.2–4.5 mm] . . . . **H. fulgidicollis**
(compare also with genitalia (1) (2) (3) & (4))

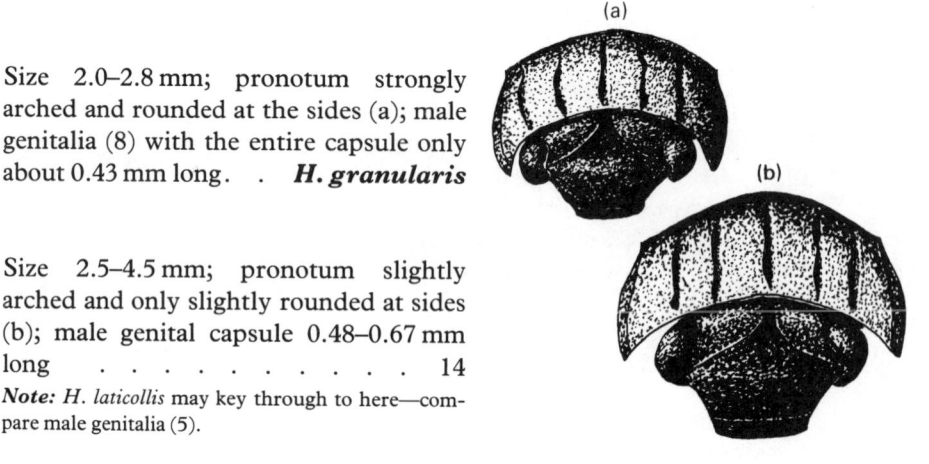

| 13 | Size 2.0–2.8 mm; pronotum strongly |
|(11)| arched and rounded at the sides (a); male genitalia (8) with the entire capsule only about 0.43 mm long. . **H. granularis** |

— Size 2.5–4.5 mm; pronotum slightly arched and only slightly rounded at sides (b); male genital capsule 0.48–0.67 mm long 14
Note: H. laticollis may key through to here—compare male genitalia (5).

| 14 | Pronotum greenish to golden with front |
|(13)| and side margins yellow; elytra yellowish, sometimes mottled, without a depression just behind the shoulders . . . . . 15 |

— Pronotum dark brown with front and side margins a dull reddish brown or darker; elytra brown or yellow, with a V-shaped depression just behind the shoulders (a) . 17

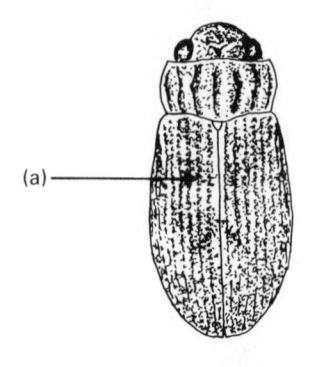

Note: all subsequent identifications rely on male genitalia.

15	RELY ON IDENTIFICATION OF MALES
(14)	
—	Tips of male parameres cut off, sloping inwards (male genitalia (9)), and the basal piece of the capsule longer than the parameres; [size 2.5–4.0 mm] **H. longitarsis**
—	Tips of the parameres pointed (male genitalia (10) and (11)), and the basal piece about the same length as, or shorter than the parameres. 16

16	Male genitalia with the tube of the aedeagus much longer than the struts (male genitalia (10)), and rather membranous, so that it may bulge; [size 2.6–3.8 mm] . . . .**H. griseus**
(15)	
—	The tube of the aedeagus about the same length as the struts and not membranous (male genitalia (11)); [size 2.5–3.5 mm] . **H. minutus**

17	RELY ON IDENTIFICATION OF MALES
(14)	
—	Male genital capsule dark brown (unless immature and soft) and large, between 0.57 and 0.70 mm long (male genitalia (12)) [size 2.6–4.0 mm]. . **H. flavipes**
—	Male genital capsule pale yellow and smaller, 0.46–0.60 (rarely 0.65) mm long; (male genitalia (13)); [size 2.7–4.3 mm] . **H. obscurus**

This key is based on Angus (1978) and its unpublished revised version (1987) by kind permission of the author. The figures of the male genitalia are redrawn from the same source.

	cpts 6–7 Last segment of palp symmetrical		cpts 8–10 Y-shaped groove narrow, parallel-sided			cpts 11–12 Wide elytral flanks		cpt 13 Small size	cpts 14–16 Paler species			cpt 17 Darker species	
ELYTRAL FLANKS W=WIDE, N=NARROW	W	W	W	W	W	W	W	N	N	N	N	N	N
GROUND COLOUR OF ELYTRA B=BROWN Y=YELLOW	B or Y	B or Y	B	B	B	B + pale spots	Y	B or Y	Y	Y	Y	B	B rarely Y
SIZE (mm) 4.0 – 3.0 – 2.0	arvernicus	brevipalpis	nanus	strigifrons	laticollis	dorsalis	fulgidicollis	granularis	longitarsis	griseus	minutus	flavipes	obscurus
Relative proportions of tube (T) + struts (S), basal piece (B) + parameres (P):	T~S B>P	T~S B~P	S>T B~P	S>T P>B	T~S B~P	T~S B~P	T~S P>B	T~S B~P	T>S B>>P	T>>S B~P	T~S B~P	S>>T B>P	S>>T P≥B
	①	②	③	④	⑤	⑥	⑦	⑧	⑨	⑩	⑪	⑫	⑬

NOTE: If male genital capsules are mounted on slides, the protrusion of the aedeagus and degree of divergence of the parameres may vary with pressure of a coverslip

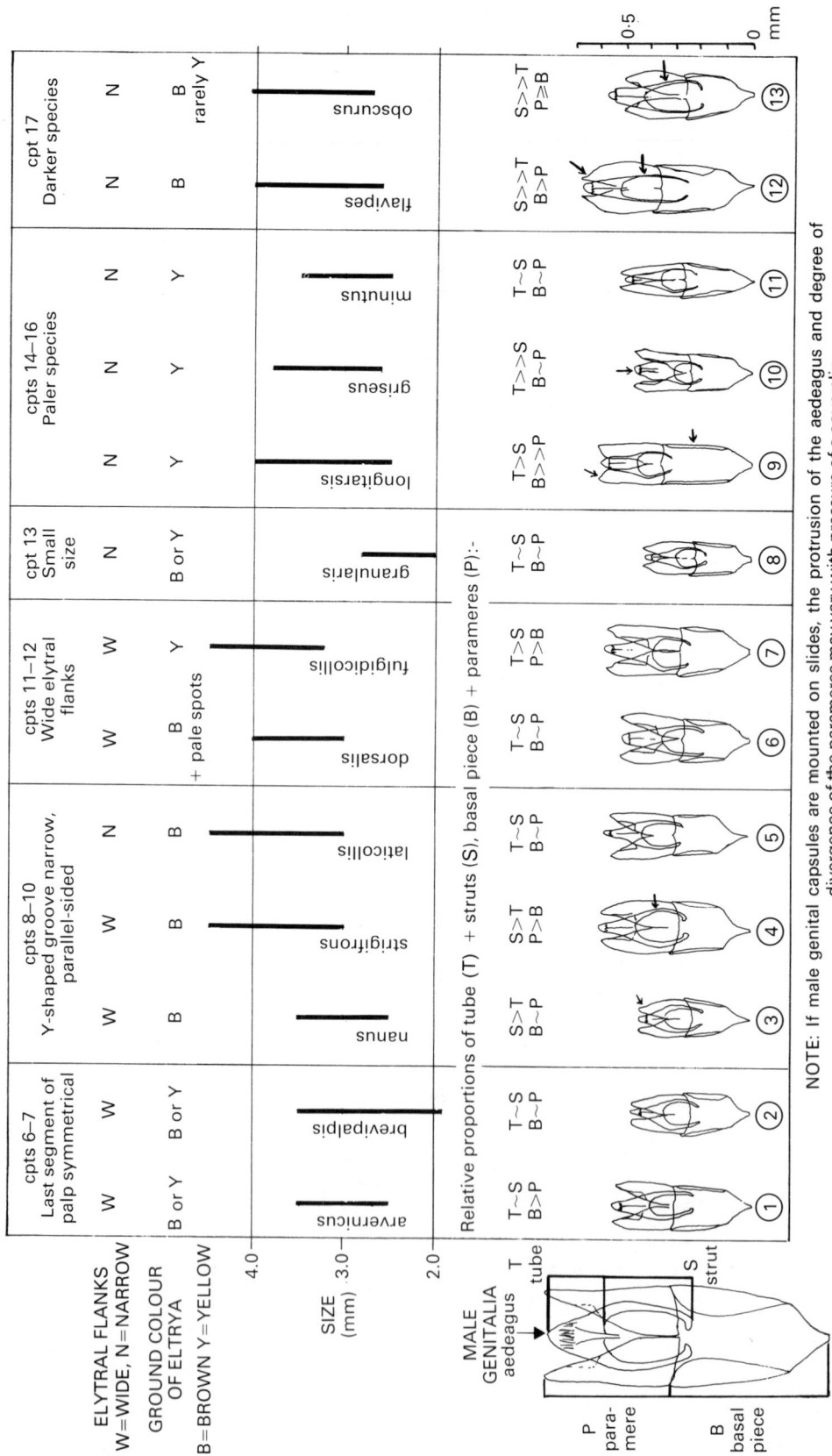

Figure 10. Tabular guide to *Helophorus* species without an extra row of punctures on the elytra. To be used in conjunction

Key G2   *Hydrochus*

Genus Characters: pronotum narrowing behind, with 5 depressions; elytra with large punctures in rows, often with metallic coloration.

Size range 2.4–4.7 mm.
6 spp.

1   Spaces between alternate rows of punctures on the elytra raised into sharp ridges, so that the rows seem to be in pairs (a).   2

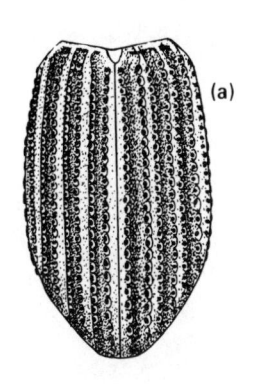

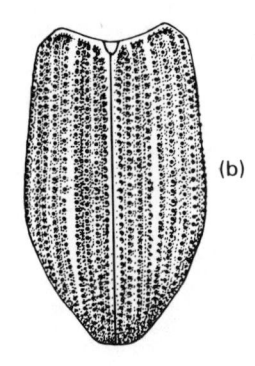

—   All spaces between rows on the elytra equally raised (b), (except for slight ridges on the front edge and near the middle of the elytra, between rows 4 & 5 from the centre, in one species) 5

2 (1)   Ridge between elytral rows 2 & 3 ending half-way back, and a ridge between rows 3 & 4 beginning at this point (a) . . . 3

—   Ridge between rows 2 & 3 running the full length of the elytra 4

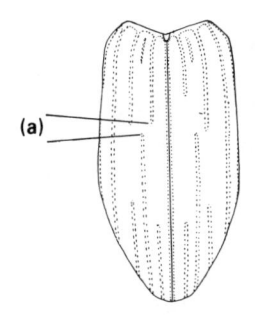

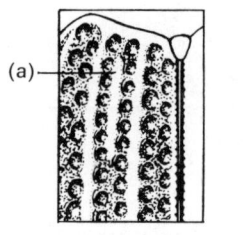

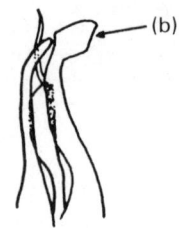

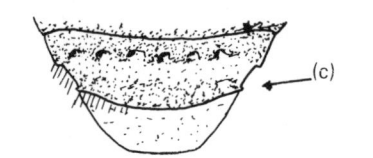

3	Space between rows 3 & 4 with a short
(2)	ridge at the front edge of the elytra, about
	same width as the neighbouring ridges (a);
	male genitalia with the tip of one paramere
	greatly enlarged (b); female with a notch
	on either side of the last sclerotised
	abdominal segment (c); [size 3.3–4.7 mm]
	H. elongatus

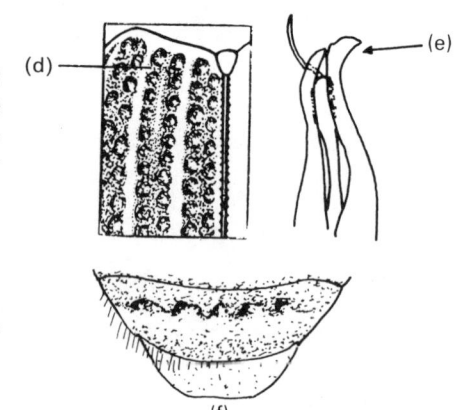

— Only a very slight narrow ridge (at most) in the position described above (d); male with tip of one paramere only slightly enlarged (e) (see note on extraction of genitalia p. 5—enter through the dorsal surface); female without notches in the sides of the last sclerotised abdominal segment (f); [size 3.6–4.0 mm] . . . .
H. ignicollis

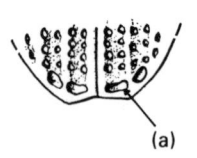

4	Body elongate, elytra almost twice as long
(2)	as the total width; hind edge of the elytra
	with large oval punctures right through
	(a), (look at rear view); [size 2.5–3.1 mm] .
	H. carinatus

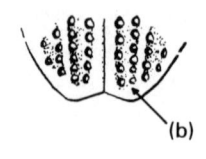

— Body short and broad, elytra about 1.5 × as long as wide; hind edge of the elytra without large punctures (b); [size 2.8–3.7 mm] . . . . ***H. brevis***

Note: A 7th species, *H. megaphallus*, has recently been recognised as British (see Berge Henegouwen, Balfour-Browne Club Newsletter *42*). It resembles *H. brevis*, except that the male genital capsule is about 1/3 the body length compared to less than 1/5 in *H. brevis*.

5	Elytra with small bumps (raised ridges) in the positions shown (a) (view at × 20 with light from the side); elytra 1.5 × as long as their total width; punctures through the hind edge of the elytra are narrow oval (b) (rear view); [size 2.4–3.0 mm] . . . . **H. nitidicollis**
(1)	

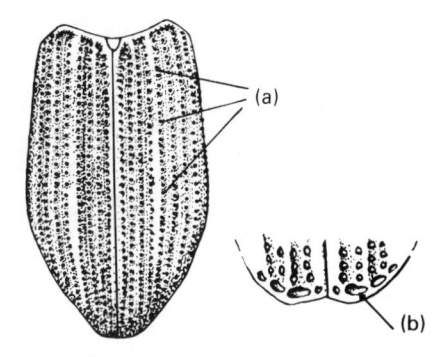

— No raised bumps on the elytra; body elongate, elytra 2 × as long as wide; punctures in the hind edge of the elytra are large and round (c); [size 3.0–4.0 mm] . . . . . **H. angustatus**

Key G3   *Cercyon*

Genus Characters: elytra with rows of large punctures; first segment of all tarsi long.
   Size range 1.3–4.4 mm.
   21 species, mostly terrestrial, in dung or rotting vegetation; 9 species characteristically found in or near water, amongst vegetation.

See Freude *et al.* (1971) for a complete key; an adapted version in English is given in the Balfour-Browne Club Newsletter **7**. The male genitalia are redrawn here, all to the same scale.

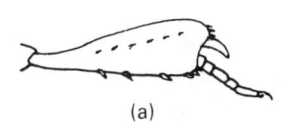

1   Front tibia cut across obliquely at the tip, and ending in a prominent spine (a); amongst seaweed; male genitalia (1); [size 2.6–3.0 mm] . . . . . **C. littoralis**

—   Front tibia rounded at the tip (b) . . 2

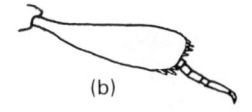

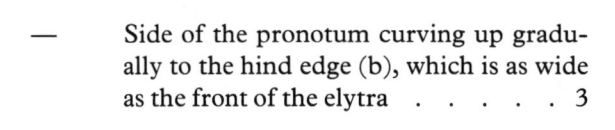

2	Side of the pronotum (viewed from the side), curving up abruptly just before the hind margin (a); viewed from above the hind edge of the pronotum slightly narrower than the front of the elytra; amongst seaweed; male genitalia (2); [size 2.2–2.4 mm] . . . . . *C. depressus*
(1)	

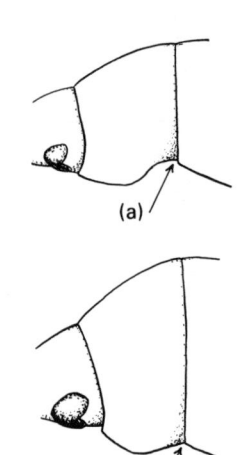

— Side of the pronotum curving up gradually to the hind edge (b), which is as wide as the front of the elytra 3

3	Elytra black with, at most, the tips of the elytra with a yellowish or reddish mark, which may extend up the side rim of each elytron in a narrow stripe (a) 4
(2)	

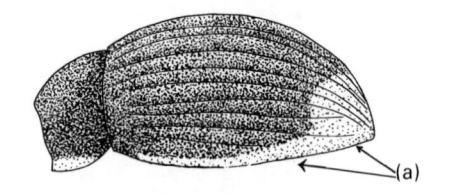

— Elytra brown, reddish or yellowish, in strong contrast to black head, with or without black markings; if elytra are predominantly black, the yellow mark at the tip extends in a broad band along the side margins (b) . . **Terrestrial species**

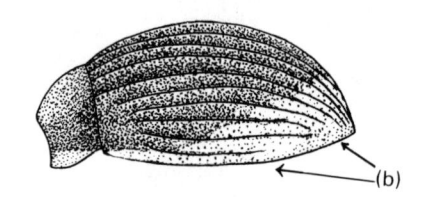

4	In side view, the elytra and pronotum form separate domes, meeting at an angle (a); male genitalia (3); [size 2.6–3.2 mm] **C. ustulatus**
(3)	

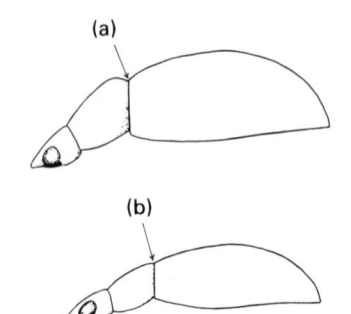

— Elytra and pronotum forming a single curve (b) 5

5	Elytra with very fine net-like reticulation, visible at × 30 on a dry beetle, between the puncture rows; elytra appearing slightly frosted in comparison with the glossy pronotum at lower magnification; size 1.6–2.3 mm 6
(4)	

— Elytra without net-like reticulation, though fine lines may be visible at × 30; elytra as glossy as pronotum; size 1.7–4.2 mm 9

6	The raised area in the midline of the underside in **front** of the **middle** coxae (mesosternal platform (a)) separated from the raised area **behind** the middle coxae by a narrow gap (b) 7
(5)	

— Mesosternal platform touches the raised area behind the middle coxae (c) . . 8

7	All rows of punctures remaining strong right up to the tip of the elytra (rear/side view at × 25) (a); elytra shining; male genitalia (4); [size 1.7–2.3 mm] . . . . ***C. granarius***
(6)	

— Punctures in the middle rows of each elytron becoming progressively weaker towards the tip, almost disappearing in the area shown (b); elytra dull; male genitalia (5); [size 1.7–2.3 mm] . ***C. tristis***

8	Mesosternal platform more than twice as long as wide (a); last segment of palps darkened; male genitalia (6); [size 1.6–2.2 mm] . . . *C. convexiusculus*
(6)	

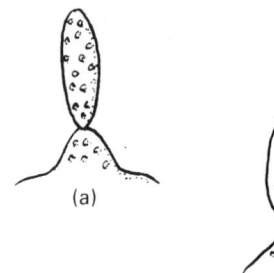

— Mesosternal platform only about 1.5 times longer than wide (b); palps uniformly reddish brown; male genitalia (7); [size 1.6–2.0 mm] . . . . . *C. sternalis*

9	Pronotum all black **Terrestrial species** (Beetles less than 2.2 mm long key out here)
(5)	

— Pronotum black with a thin paler stripe along the side margin, or a yellowish spot at the front corner 10

10	Pale side borders of the elytra extending almost to the front edge (a) (side view); mesosternal platform (see couplet 6) nearly 3 times as long as broad; male genitalia (8); [size 2.2–3.0 mm] . . *C. marinus*
(9)	

— Pale side borders extending only half way up the elytra (b) mesosternal platform less than twice as long as wide; male genitalia (9); [size 2.2–3.0 mm] *C. bifenestratus*

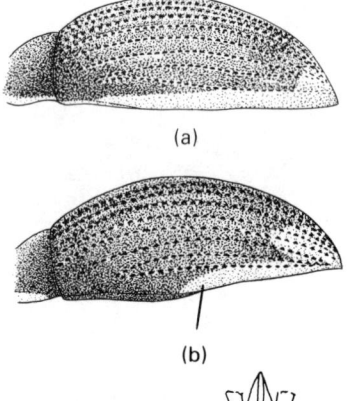

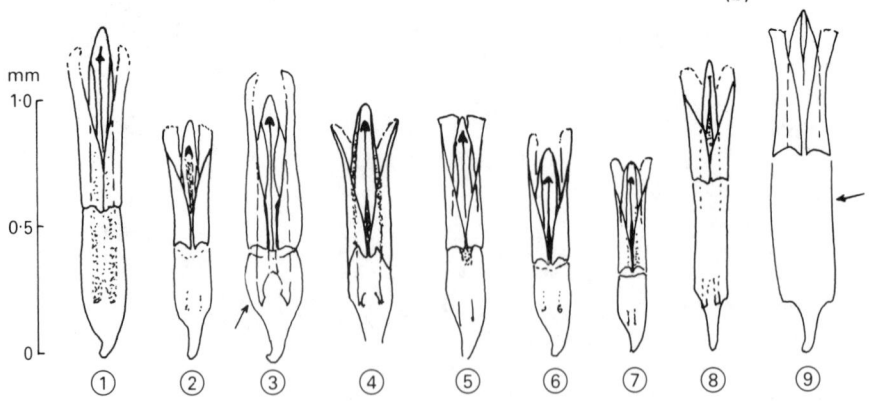

Cercyon ♂ genitalia
(redrawn from Freude et al., 1971)

(1) *C. littoralis*; (2) *C. depressus*; (3) *C. ustulatus*; (4) *C. granarius*; (5) *C. tristis*; (6) *C. convexiusculus*; (7) *C. sternalis*; (8) *C. marinus*; (9) *C. bifenestratus*.

Key G4 *Berosus*

Genus Characters: elytra with 10 longitudinal furrows formed of pits; middle and hind tibiae with a fringe of pale hairs; may squeak if alarmed.
Size range: 3.5–5.5 mm.
4 spp.

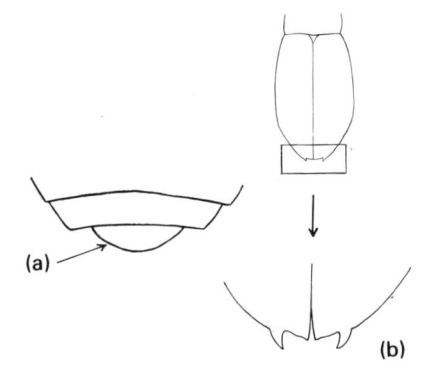

1   Last abdominal segment (underside) with smoothly rounded margin (a); conspicuous spine present near the tip of each elytron (b) (tilt the end of the beetle up slightly to view from above, or view from below); head yellow; elytra greenish; [size 4.5–5.5 mm] . . . . . ***B. spinosus***

—   Last abdominal segment with indentation at the tip (c); spines on elytra very inconspicuous or absent; head dark, metallic   2

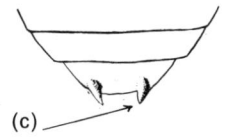

2 (1)   Dark metallic coloration on the pronotum restricted to two longitudinal stripes (a) (about the same width as the front tibia); keel between the front and middle coxae (mesosternal keel) with profile as shown (b) [size c. 5.5 mm] . . ***B. signaticollis***

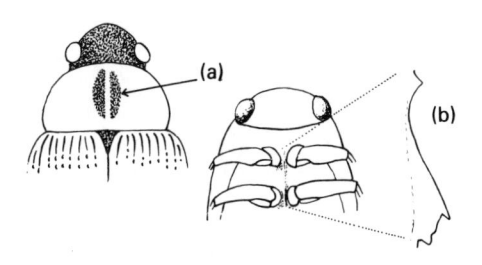

—   Dark metallic coloration on the pronotum in large patches (c) (wider than the front tibia); size less than 5.0 mm 3

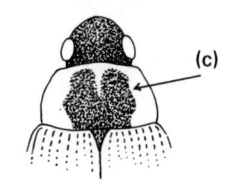

3 (2)	Mesosternal keel rising to a high ridge, with profile as shown (a); areas between elytral grooves raised into rounded ridges (b) (view at ×15 or more, light from the side). . . . . **B. luridus**	
—	Mesosternal keel very weak, without a high ridge (c); areas between the elytral grooves flat (d) . . . . . **B. affinis**	

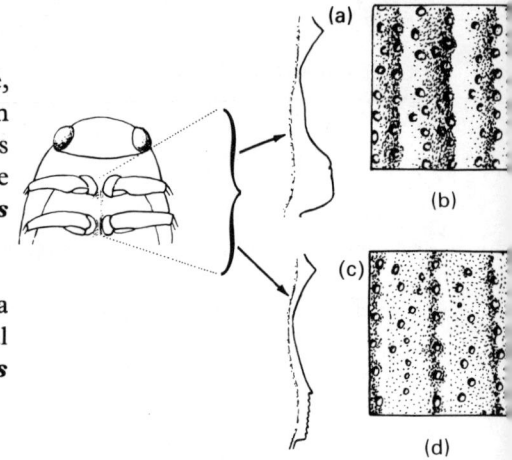

Key G5   *Helochares*

Genus Characters: palps about twice as long as antennae; no groove on either side of the midline of the elytra; front claws bent at right angles; females carry eggs beneath the abdomen, in strands of silk, in early summer.
Size range 4.5–6.5 mm.
3 spp.

1   Head pale with, at most, the hind edge darkened, and the same colour as the pronotum; elytra yellow to reddish and shining, with very fine "pinpoint" punctures (a) and two or three lines of stronger punctures on each elytron; male genitalia (1); [size 4.5–6.5 mm, males smaller than females] . . . . **H. lividus**

—   Head dark; elytra red or dark, with strong puncturation, the rows of larger punctures less apparent 2

2   Head all dark, or reddish with a dark hind
(1)   margin; punctures on elytra large and very dense, the space between the punctures much less than their diameter (b); male genitalia (2) . . . . . **H. obscurus**

—   Head all black, or dark red with a black central band; punctures on the elytra less dense, the space between punctures about equal to their diameter (c); male genitalia (3) . . . . **H. punctatus**

Note: H. obscurus (Müll.) is a species new to Britain, the species previously known as *H. obscurus* Sharp is *H. punctatus*.

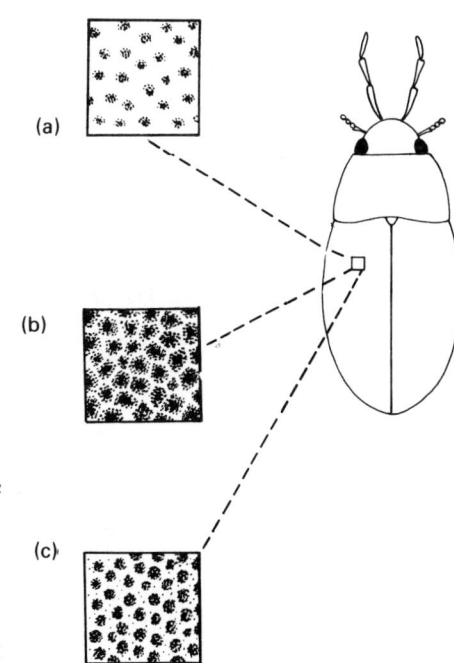

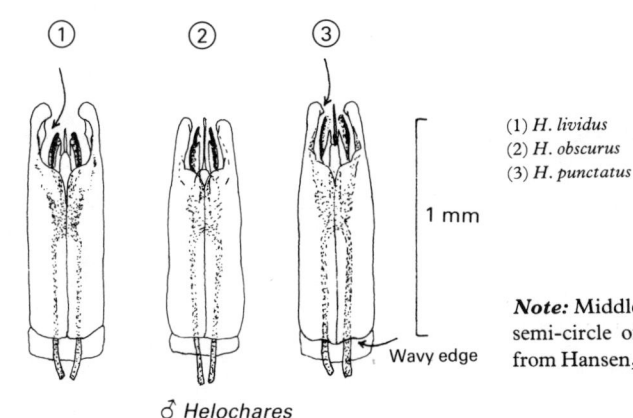

(1) *H. lividus*
(2) *H. obscurus*
(3) *H. punctatus*

1 mm

Wavy edge

♂ *Helochares*

Note: Middle part may be extruded, in which case a semi-circle of strong spines is revealed. Redrawn from Hansen, 1987.

Key G6   *Enochrus*

Genus Characters: elytra with large punctures not obviously in rows; palps up to twice as long as the antennae, with the longest segment curving outwards.
Size range 3.7–7.0 mm.
10 spp. (see Hansen, 1987).

Note: Some species are most reliably distinguished by features of the male genitalia, recognisable from the protruding tips; see note on the extraction of genitalia on p. 5.

1   Head, pronotum and elytra all greenish yellow or brown; palps all yellow; [size 5.0–6.5 mm] . . . . . . ***E. bicolor***
    Note: Some females have small dark marks between the eyes.

—   Head black, at least between the eyes; palps may have dark markings . . . 2

2   Last segment of the palps same length as
(1) the previous one (a); elytra and pronotum orange or yellow; [size 5.0–5.5 mm] . .
                            E. melanocephalus

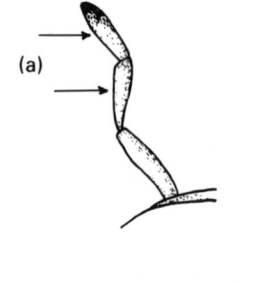

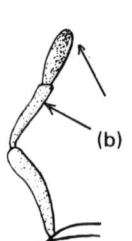

—   Last segment of the palps shorter than the previous one (b); elytra and pronotum brown 3

3   Size 4.8 mm or larger; last segment of
(2) abdomen (underside) smoothly rounded at tip 4

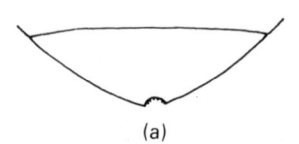

—   Size 4.6 mm or less; last abdominal segment with a small semicircular bay at the tip (× 20) (a) 8

4 (3)	First visible segment of the palps dark in basal half (a); [size 6.0–7.0 mm] . . . . **E. testaceus**	(a)
—	First visible segment of the palps not darkened 5	
5 (4)	Traces of two or three longitudinal rows of sparse punctures on the elytra (view at × 20; light from the side, on a **dry beetle**) (a); male with long, very sharply bent front tarsal claws with a very large tooth at the base, much larger than on the middle or hind tarsal claws (b) and genitalia with a broad middle part (aedeagus) (c) . . 6	(a) (b) (c)
—	No trace of rows on the elytra; male front tarsal claws short, less sharply bent (d), and genitalia with slender aedeagus (e); female, but not the male, with tip of the palp darkened (see also next couplet); [size 4.8–5.5 mm] . . . . **E. ochropterus**	(d) (e)
6 (5)	Tip of the last segment of palp dark in both sexes; in freshwater 7	
—	Palps uniformly pale; in brackish water; [size 5.0–5.8 mm] . . **E. halophilus**	

7 (6)	Pronotum with a central black spot and four small spots around this arranged in a square (a); labrum of male yellow-red (b), of female, black; [size 5.0–5.7 mm] . . . ***E. quadripunctatus***	
—	Central black spot of pronotum very large, usually obscuring some or all of the smaller spots (c); labrum of both sexes black (d); [size 5.0–5.5 mm] . . . . . ***E. fuscipennis*** *Note:* males have very strongly curved front tarsal claws, those of females are only moderately curved.	
8 (3)	Obvious yellow or brownish triangular mark in front of the eyes (a); outer parts of the male genitalia (parameres) only slightly curved and rounded at the tip (b) 9	
—	Faint yellow mark may be present in front of the eyes; male parameres outward-pointing at the tip (c); [size 3.2–3.7 mm] . ***E. affinis***	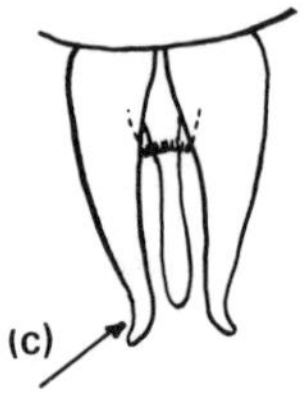
9 (8)	Middle part of the male genitalia broad, outer parts (parameres) straight or curved inwards (a); dark band down the last two-thirds of the mid-line of the elytra; [size 3.5–4.6 mm] . . . . ***E. coarctatus***	
—	Middle part of the male genitalia slender; parameres slightly divergent (b); thin dark line down the centre of the elytra, no broader band towards the rear; [size 3.4–4.1 mm] . . . . . . ***E. isotae***	

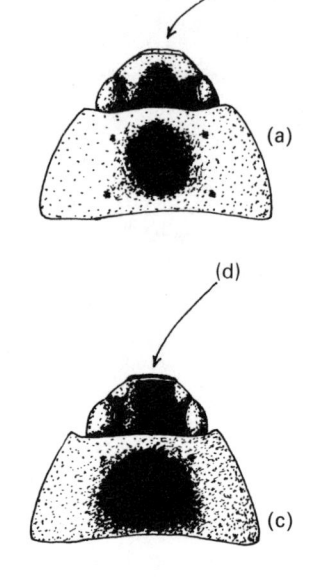

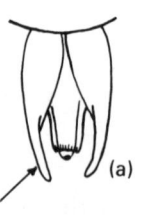

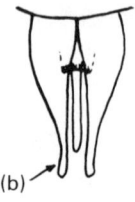

Key G7　　　　　　　　　*Laccobius*

Genus Characters: elytra with strong punctures usually marked by pigment spots; pronotum with dark central area.
Size range 2.5–4.0 mm.
9 spp.

Note: for several species, certain identification is possible only by examination of male secondary sexual characters and genitalia.

Males differ from females in having the 2nd and 3rd segments of the front tarsi broader than their neighbours and segment 3 slightly darkened. See note on extraction of genitalia, p. 5. Male genitalia are redrawn from Gentili, 1977.

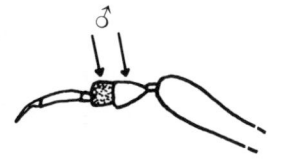

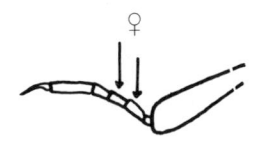

1	Head, pronotum and elytra purple metallic; [size 3.5–4.0 mm]. . . . . . . 　　　　*L. striatulus* v. *purpurascens*	
—	Elytra blotched yellow and grey, appearing lighter than the centre of the pronotum; size 2.5–4.0 mm 2	
2 (1)	Size 2.5–3.0 mm; punctures on the elytra in definite straight rows with hardly any out of line (a) (look at the area about halfway down, 3 to 6 rows from the midline); males without "goggles" (see below) . 3	
—	Size 3.5–4.0 mm; elytral punctures not all strictly in line (b), especially in the area described above; males with glassy "goggles" on the underside of the labrum (view from underneath), either round (c), oval (d) or narrow crescents (e) . . . 4	

Note: a 9th species which might be discovered in Britain, *L. obscuratus* Rottenberg, has no goggles in the male, has a large pronotal patch (like *L. atratus*), irregular puncture rows and is 2.8–3.7 mm long. Male genitalia (9).

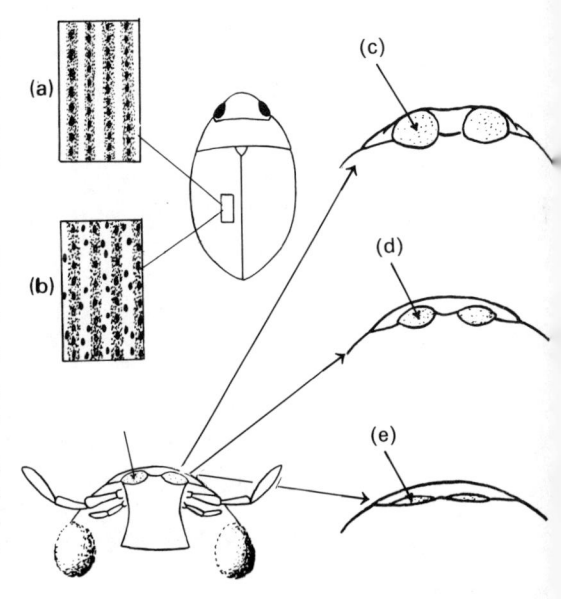

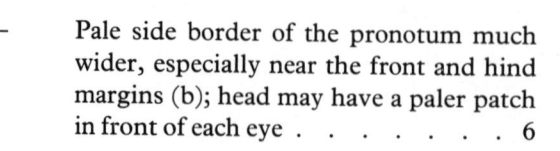

3 (2)	Pronotum with minute reticulation, the individual meshes visible between the punctures at × 40 (light from the side, dry beetle); tip of each elytron usually without a conspicuous pale patch; antennal club dark brown, much darker than rest of antenna; male genitalia (1) ***L. minutus***	
—	Pronotum without reticulation, smooth and glossy between punctures; tip of each elytron usually with a conspicuous yellow patch (a); antennal club yellowish, the same as the rest of the antenna; male genitalia (2) . . . . .***L. biguttatus***	(a)
4 (2)	Dark patch on the pronotum narrow, never approaching the hind or side margins (a); elytra pale; male with circular goggles; male genitalia (3) ***L. simulatrix***	(a)
—	Dark patch on pronotum extensive and reaching the **hind** margin (b); male with circular, oval or crescent-shaped goggles (see cpt 2) 5	(b)
5 (4)	Pale side border of the pronotum narrow, no wider than the width of an eye except, perhaps, near the hind margin (a); head all black; elytra dark; [size 3.5–4.0 mm] male goggles oval; genitalia (4) . . ***L. atratus***	(a)
—	Pale side border of the pronotum much wider, especially near the front and hind margins (b); head may have a paler patch in front of each eye 6	(b)

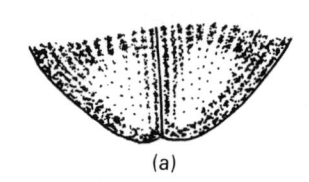

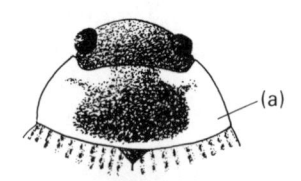

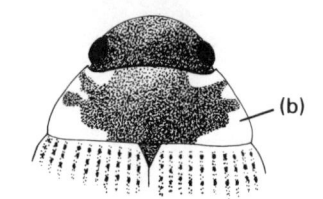

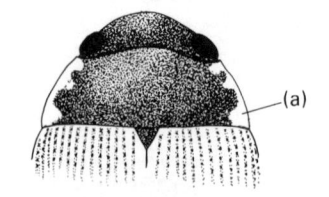

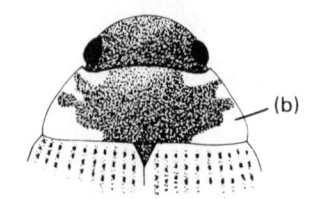

6	Pronotum with minute reticulation between the punctures, with the mesh-like pattern distinct at ×40 (light from side, dry beetle); labium (underside of head between eyes) with clearly defined, well-spaced pits (a) (×40); male with narrow crescent-shaped goggles (b); male genitalia (5) . . . . **L. bipunctatus**
(5)	

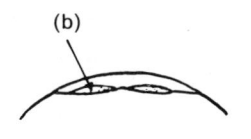

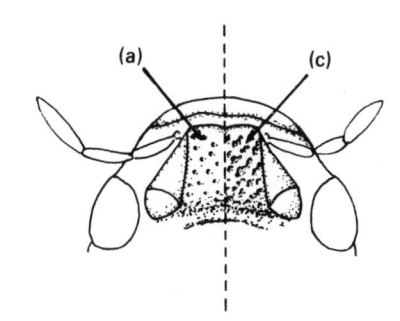

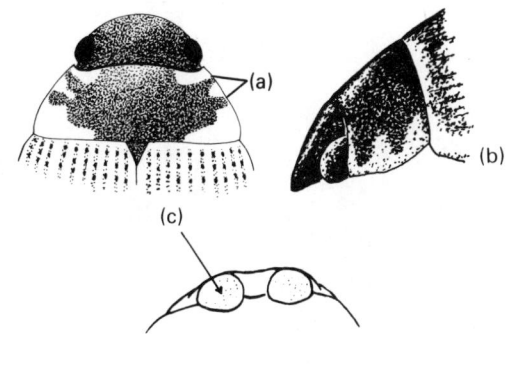

—	Pronotum either very smooth and glossy between the punctures, with no sign of reticulation, or with a slightly "frosted" appearance on some areas, but no distinct mesh pattern visible even at ×40 (if pronotum is "frosted", the labium is very densely pitted (c); if pronotum is glossy, labium may be like either (a) or (c)); male goggles circular or oval. 7

7	Dark patch on pronotum broad, almost reaching the **side** margin in places, seen from above (a), side view as (b); male goggles virtually circular (c); male genitalia (6). . . . . . .**L. atrocephalus**
(6)	

—	Dark patch on pronotum narrow, well clear of reaching the side margin (d) (e); male goggles oval (f) 8

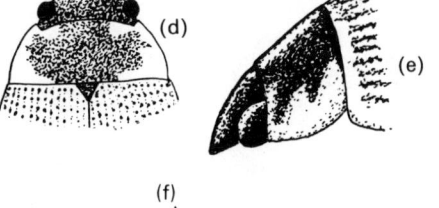

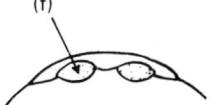

9 (8)	Base of the middle femur of males with a tuft of yellow hairs visible at ×20 (a); body round oval, elytra hardly longer than wide (b); elytra sometimes pale but usually dark with paler lines; head usually with a yellow patch in front of each eye; male genitalia (7) . . . . **L. striatulus**	
—	No tuft of hairs on the base of the middle femur of males at ×20; body long oval, elytra 1.25× as long as the total width (c); elytra pale with dark lines; head with, at most, rather indistinct paler patch in front of each eye; male genitalia (8) **L. sinuatus**	

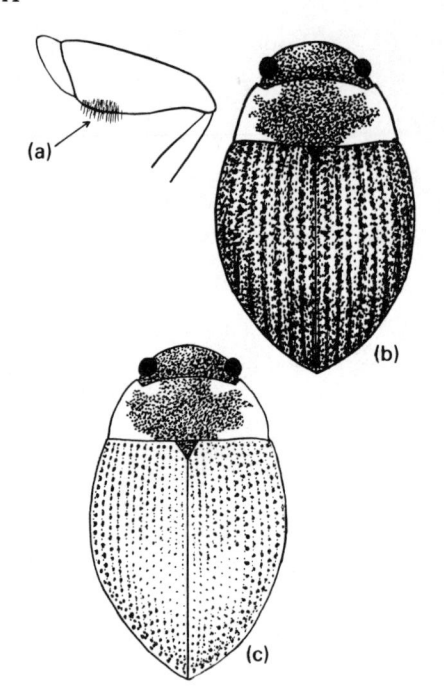

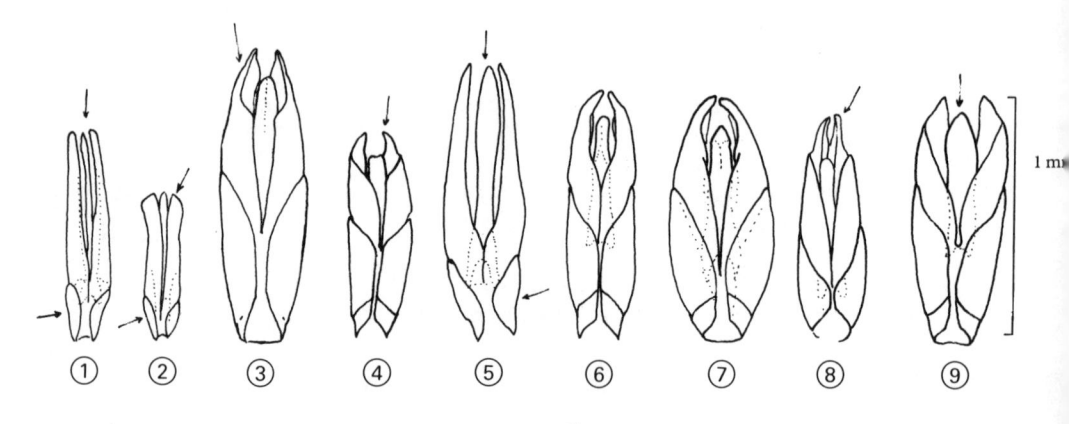

Laccobius ♂ genitalia

(1) *L. minutus*
(2) *L. biguttatus*
(3) *L. simulatrix*
(4) *L. atratus*
(5) *L. bipunctatus*
(6) *L. atrocephalus*
(7) *L. striatulus*
(8) *L. sinuatus*
(9) *L. obscuratus*

Key G8   *Paracymus* and *Anacaena*

Genera Characters: small, domed beetles; palps about equal to antennae; punctures scattered all over elytra and pronotum.
Size range 2.0–3.0 mm.
Paracymus, 2 spp.
Anacaena, 4 spp. *(See Berge Henegouwen, 1986)*

1 Hind femora with felty hairs over most of the underside, appearing the same texture as the abdominal segments (use light from side and above, dry beetle); elytra without a blue or green metallic sheen . . . . *Anacaena* . . 2

— Hind femora with the ventral face appearing shiny in relation to the abdominal segments (a); elytra with blue or green metallic sheen   . . *Paracymus* . . 5   (a)

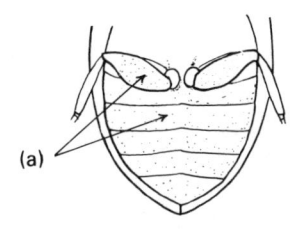

2 (1) Felty patch on the hind femur covering virtually the entire ventral face, extending more than 3/4 way along the hind margin (a); pronotum often with a dark patch in the centre, and another on either side; head usually with a pale patch in front of each eye (see cpt 4)   ***Anacaena limbata***

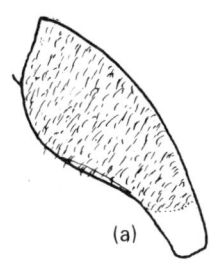

— Felty patch on hind femur "cut away" on hind margin, not reaching more than half way along it (b); pronotum with or without darker areas; head with or without pale spots 3

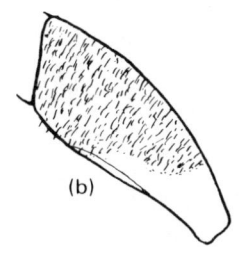

3 (2)	Ridge present, rising to a horn (a) in the midline between the front and middle coxae (shine light down between the coxae, view at × 30 or more); elytra brown 4	
—	No ridge or horn present between the front and middle coxae although a slight bump may be detectable; colour black unless immature [size 2.5–3.0 mm] . . . ***Anacaena globulus***	
4 (3)	Large yellow patch present in front of each eye (a); palps yellow with the end half of the last segment darker (b); pronotum predominantly pale [size 2.0 mm] . . . ***Anacaena bipustulata***	
—	No pale patch in front of each eye; palps brown with the last segment very dark (c); pronotum predominantly black [size c. 2.5 mm] . . . ***Anacaena lutescens***	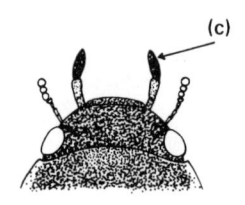
5 (1)	Legs and palps red; [size c. 3.0 mm] . . . ***Paracymus aeneus***	
—	Legs dark, palps almost black; [size c. 3.0 mm] . . . ***Paracymus scutellaris***	

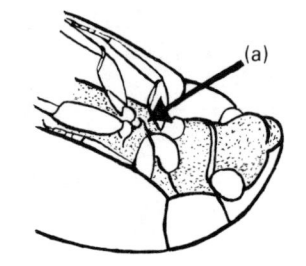

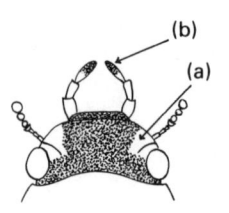

Key G9 *Limnebius*

Genus Characters: domed beetles less than 3 mm long; elytra truncate.
    5 spp.

Note: male *Limnebius* have six sclerotised abdominal segments visible on the underside; females have seven.

1   Hind tibia with sharp constriction near joint with the femur (a) [size 2.5 mm] . . male ***L. truncatellus***

—   Hind tibia slender throughout its length (b) 2

2 (1)   Penultimate segment of the palps swollen, at its widest point about twice the width of the end segment (a); elytra brown, paler than the pronotum [size 2.0–2.5 mm] . . male ***L. papposus***

—   Penultimate segment of the palps little wider than the end segment (b) . . . 3

3 (2)   Labium glossy, brown, with a wide central groove (a) (underside of the head, between the bases of the antennae); elytra brown, paler than the pronotum [size 2.0 mm] . female ***L. papposus***

—   Labium dull, black, and flat or with only a slight groove (b); mature beetle black . 4

4 (3)   Size 2.0 mm 5

—   Size 1.5 mm or less 6

5	Elytra with punctures obvious at ×20; front of the head (at the junction with the labrum) almost as wide as distance between the eyes (a). . . . .
(4)	female *L. truncatellus*

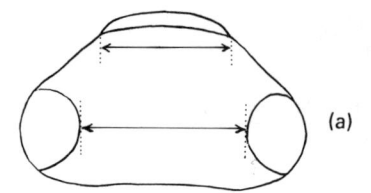

—	Elytra with fine punctures, scarcely visible at ×20; width across the front of the head much less than the distance between the eyes (b); male genitalia (1) (see next couplet) . . . . *L. crinifer*

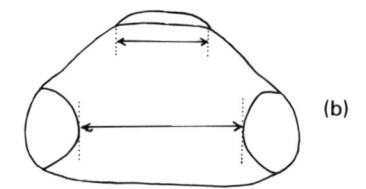

6	Size 1.5 mm; pronotum and elytra with fine punctures just visible at ×20; male genitalia (2) . . . . . . *L. nitidus*
(4)	

—	Size less than 1.0 mm; pronotum and elytra with very fine punctures, not visible at ×20; male genitalia (3) . . *L. aluta*

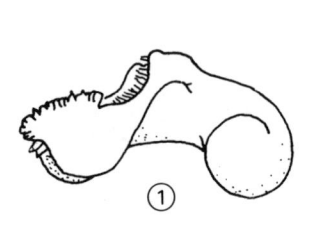

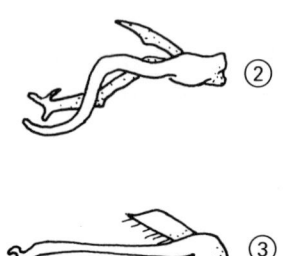

Limnebius male genitalia (redrawn from Carr, 1984a).
(1) *L. crinifer*
(2) *L. nitidus*
(3) *L. aluta*

Key G10   *Hydraena*

Genus Characters: palps very long (a); elytra with punctures in rows; labrum cleft.
    Size range 1.0–2.5 mm.
    10 spp.

Note: Some secondary sexual characteristics and male genitalia are referred to in the Key; males can be recognised by the 6 sclerotised abdominal segments (b); females have 7 (c). See note on extraction of genitalia, p. 5. Male genitalia are redrawn from Foster, 1979.

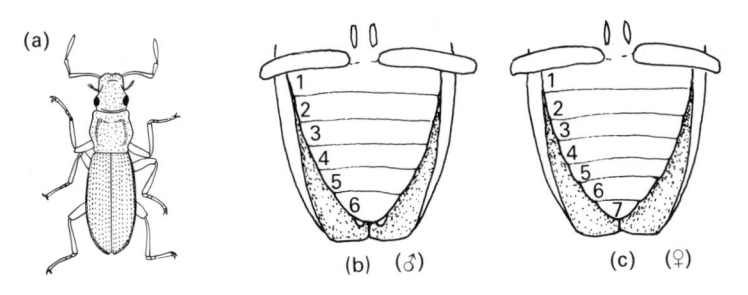

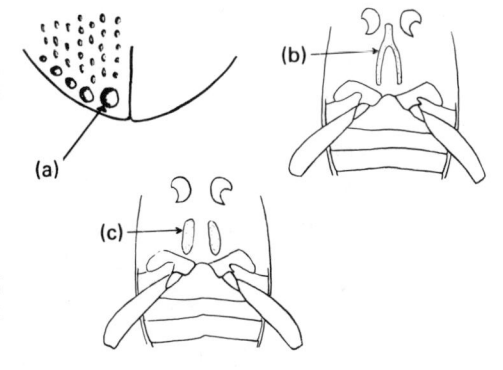

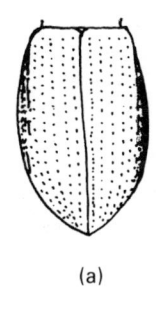

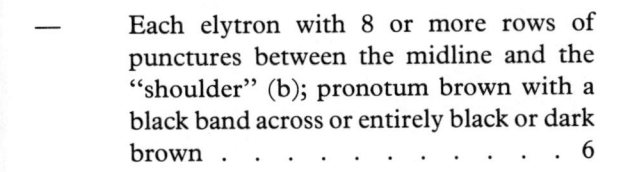

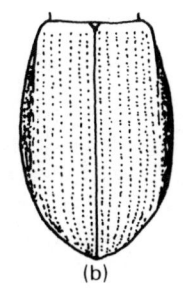

1	Tip of the elytra with large holes, twice as large as the punctures in rows (a) (rear view); a Y-shaped ridge on the underside of the thorax between the middle and hind coxae (b) [size c. 1.5 mm] . **H. testacea**	
—	No large holes at the tip of the elytra; underside of the thorax with two shining patches or narrow parallel ridges (c) . 2	
2 (1)	Each elytron with 6 or less rows of punctures between the midline and the "shoulder" (a) (× 20, light from the side); pronotum brown, with or without dark band across 3	
—	Each elytron with 8 or more rows of punctures between the midline and the "shoulder" (b); pronotum brown with a black band across or entirely black or dark brown 6	

| 3 | Size 2.0–2.4 mm; elytra elongate, more than 1.75 times as long as the total width (a); females with an indentation in the tip of the elytra (b) (view from rear—see note at top of Key, p. 131) . . **H. gracilis** |
|(2)| |

| — | Size 1.0–1.75 mm; elytra less than 1.75 times as long as broad (c); neither sex with an indentation in the tip of the elytra . 4 |

| 4 | Elytra rather flattened and well rounded at the sides, less than 1.5 times as long as broad (a); male with bulge in the hind tibia near the junction with the tarsus ($\times 20$) (b), and a tooth on the last segment of the palps (c); male genitalia with a short flagellum on the aedeagus and short parameres (d); [size 1.5–1.75 mm] . . . **H. pygmaea** |
|(3)| |

| — | Elytra well arched and not particularly rounded at the sides, more than 1.5 times as long as broad (e); male without bulge on hind tibia . . . . 5 |

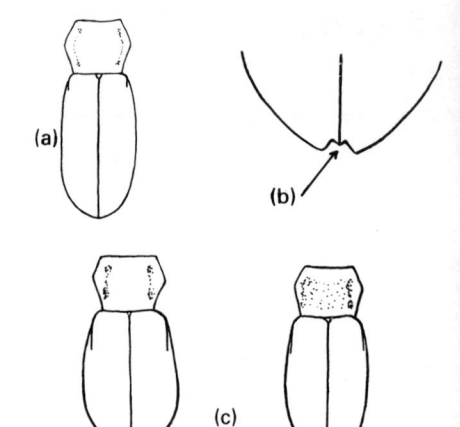

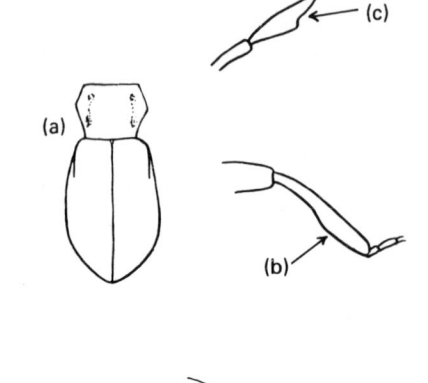

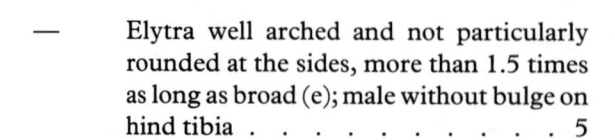

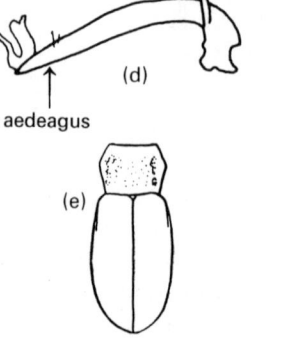

A Key to the Adults of British Water Beetles

5 (4)	Elytra cut off squarely at the tip (a) (rear view); male with a tooth on the last segment of the palps (× 40) (b); male genitalia with a long flagellum and with apparently only one paramere (c); [size c. 1.0 mm] . . **H. minutissima**	
—	Elytra evenly rounded at the tip (d); pronotum with a dark band across; male without a tooth on the palps; male genitalia with no flagellum but with long parameres (e); [size 1.2–1.5 mm] . . **H. pulchella**	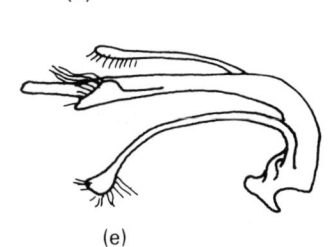
6 (2)	Side margins of pronotum sloping sharply inwards in rear half but much less so in front half (a); front margin of the pronotum strongly curved inwards (b); pronotum dull brown with a darker band across; [size c. 1.5 mm]. . **H. palustris**	
—	Side margins of pronotum sloping in about equally in the front and rear (c); front margin of pronotum only very slightly concave (d); pronotum black or dark brown, shining 7	

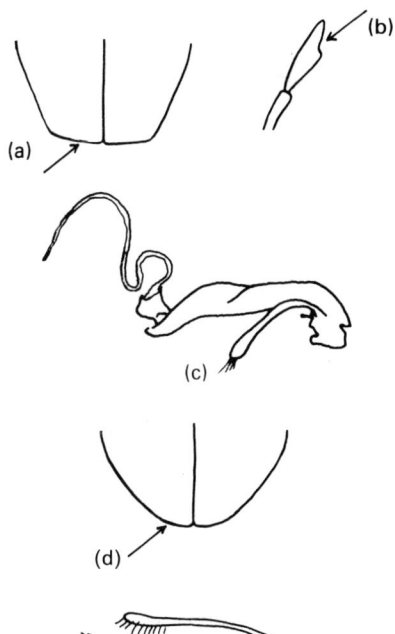

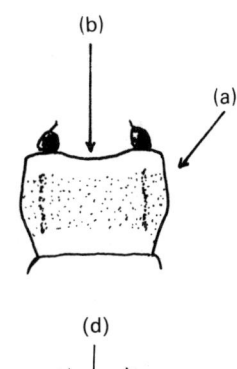

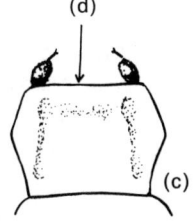

7 (6)	Hind tibia with a bulge near the joint with the tarsus (× 20) (a) [size 1.5–2.0 mm]. . male ***H. rufipes***	(a)
—	Hind tibia without a bulge near the joint with the tarsus 8	
8 (7)	Elytra well rounded at the sides, less than 1.5 times as long as broad (a); beetle black [size 1.5–2.0 mm] . . . . ***H. nigrita***	(a)
—	Elytra almost parallel-sided, almost 1.75 times as long as broad (b); mature beetle brown or black 9	(b)
9 (8)	Last segment of the palps with a definite tooth (a) (b) (× 40) ***males*** (see note at top of Key) . . . 10	(a) (b)
—	Last segment of the palps with, at most, a slight swelling towards the middle (c) . . ***females*** of ***H. rufipes, H. britteni*** & ***H. riparia.*** Rely on identification of males	(c)

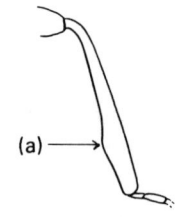

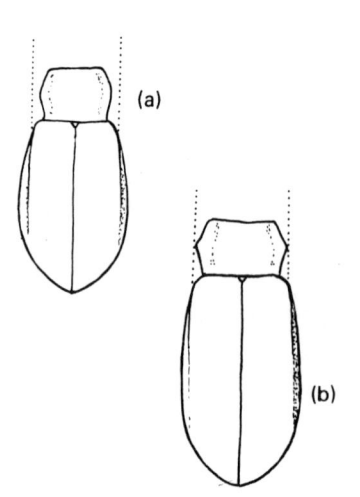

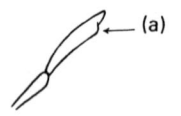

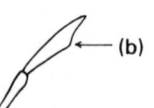

10 (9)	Tooth on the palps in the end 1/4 of the last segment (a) (× 40); male genitalia with flagellum shorter than aedeagus (b) [size 1.75–2.0 mm]. . . . male ***H. britteni***	
—	Tooth on the palps 1/3 way along the last segment (c); male genitalia with flagellum and aedeagus about equal in length (d); [size 2.0–2.5 mm] . . male ***H. riparia***	

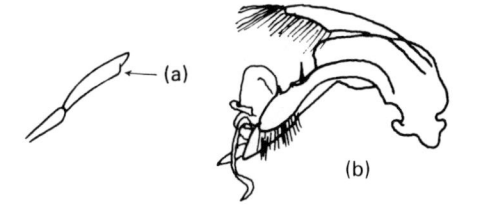

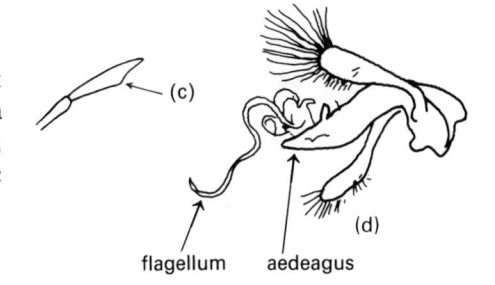

flagellum   aedeagus

Key G11   *Ochthebius*

Genus Characters: Pronotum narrowing behind; palps shorter than head, with the last segment shorter than the previous one; elytra with longitudinal rows of large punctures.
Size range 1.0–2.8 mm.
15 spp.

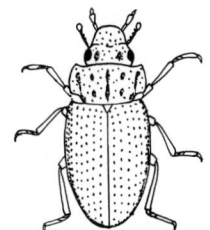

General form of pronotum and terms used in key. View at × 20, with the light from above and the side, on a **dry** beetle.

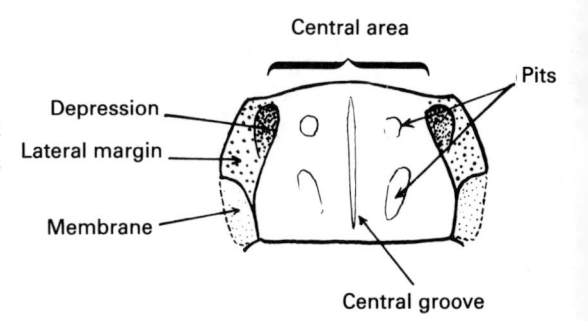

Central area
Pits
Depression
Lateral margin
Membrane
Central groove

1	Pronotum with deep transverse grooves right across and a deep central groove (a); pronotum black and shining, without punctures [size 1.0 mm] . ***O. exaratus***
—	Pronotum without grooves right across, though grooves may be present on the central area (b), (c); central area of the pronotum with punctures; size 1.25 mm or more. 2
2 (1)	Pronotum with two wide transverse depressions (which appear matt in contrast to surrounding areas) and no **continuous** midline groove (a) . . . 3
—	Pronotum without such transverse depressions (though narrow pits or faint grooves may be present) and with a midline groove (b) 6
3 (2)	Underside with a shiny patch (×30) just in front of the hind coxae (a); labrum with a small semi-circular notch (tilt up the head and view face-on at ×40) (b); elytra dark, bronzy, [size 1.5–1.7 mm] . . ***O. pusillus***
—	Underside entirely felty between the middle and hind coxae; labrum without a notch, with a smoothly rounded front margin 4

4 (3)	Elytra brown, often pale, much paler than the head; [size 1.8–2.1 mm] ***O. marinus***	
—	Elytra very dark brown or black, with or without metallic reflections, hardly, if at all, paler than the head 5	
5 (4)	Size 1.7–2.0 mm; pronotum widest between 1/3 and 1/2 way back from the front margin (a) . . . ***O. lenensis***	(a)
—	Size 1.4–1.6 mm; pronotum widest about 1/4 way back from the front (b) ***O. viridis*** **Note:** look directly down on top of the pronotum; if the pronotum is tilted down at the front, foreshortening will occur.	(b)
6 (2)	Pronotum with two pairs of well-defined pits, either side of the mid-line groove (a) (these pits, where present, are as well-defined as the lateral depression–see introductory figure; they are just visible at × 10 with side lighting and are very obvious at × 20) 7	(a)
—	Pronotum without such obvious pits on either side of the midline, though faint marks in these positions may be discernable above × 20 magnification (b) . 13 (If in doubt as to whether definite pits are visible at × 20, take this lead).	(b)

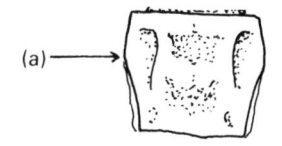

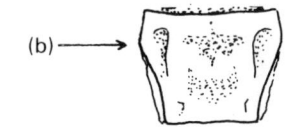

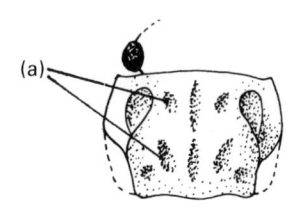

7 (6)	Elytra much wider than the pronotum and almost as broad as long; the sides of the elytra splaying out at a shallow angle and clearly visible well outside the line of the shoulder (a); pronotum narrowing half way back; male pronotum highly domed; second antennal segment cup-shaped in both sexes ($\times 30$) (b) [size 1.2–2.3 mm] . ***O. exsculptus***	
—	Elytra only a little wider than the pronotum and distinctly longer than broad, with the sides almost vertical behind the shoulder and only narrowly visible from above (c); pronotum narrowing $\frac{1}{4}$ or $\frac{1}{2}$ way back or in the last quarter; pronotum never domed; second antennal segment narrowing at both ends (d) 8	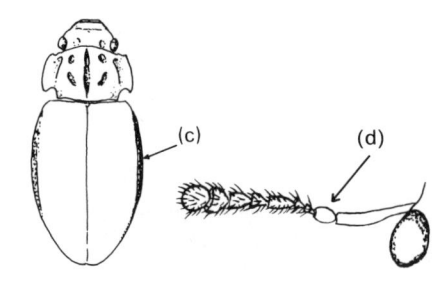
8 (7)	Pronotum narrowing about 1/4 to 1/2 of the way from the front (a); membrane longer than wide 9	
—	Pronotum narrowing in the last quarter in a scooped-out bay (b); membrane as wide as long 11	

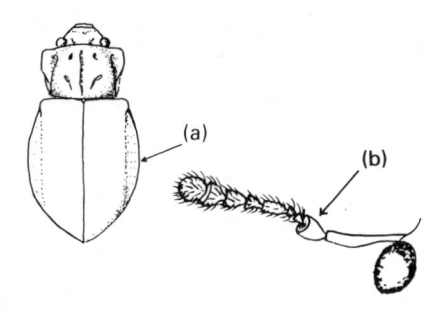

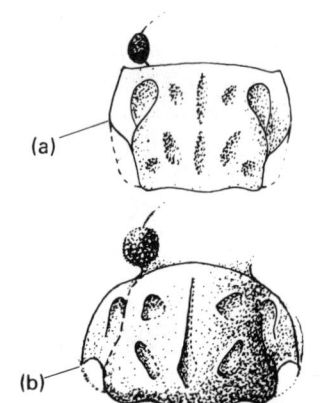

Note: look directly down on top of the pronotum; if the pronotum is tilted down at the front, foreshortening will occur.

9 (8)	Size 2.5–2.8 mm; beetle black and shining with long white hairs (these may fall out) . . . . ***O. punctatus***
—	Size 1.3–1.8 mm; beetle black and shining with hairs **or** matt without long white hairs. 10

10 (9)	Areas between the pits on the pronotum heavily punctured, appearing granular and matt (×20 or more, light from the side); elytra with obvious grooves along the rows of punctures; no long white hairs; [size 1.3–1.5 mm] . . . . **O. poweri**	
—	Areas between the pits on the pronotum with only very sparse, minute punctures, appearing smooth and glossy; elytra without longitudinal grooves; long hairs may be present; [size 1.5–1.8 mm] **O. nanus**	
11 (8)	Elytra broadest at about 2/3 the way back and narrowing abruptly behind this point (a); strips between the rows of punctures on elytra convex, like "ridge and furrow" (b) (view at ×20 with light from the side on a **dry** beetle); [size 1.5–1.8 mm] . . . **O. bicolon**	
—	Elytra broadest at about 1/2 the way back and narrowing less abruptly behind (c); strips between puncture rows virtually flat (d) 12	
12 (11)	Lateral margins of the pronotum hardly wider than an eye, the sides of the pronotum not extending beyond the shoulders of the elytra (a); labrum with a straight front margin (b), (raise the beetle's head and look at ×30 or more, face-on); beetle deep red-brown [size 1.8–2.1 mm] . **O. dilatatus**	
—	Lateral margins of the pronotum much wider than an eye, the sides of the pronotum extending beyond the shoulders of the elytra (c); labrum strongly concave on the front margin (d); [size 1.5–1.9 mm] **O. auriculatus**	

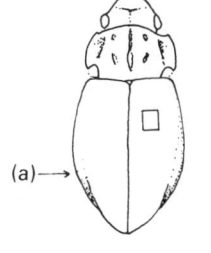

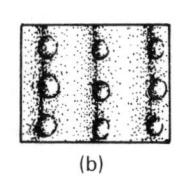

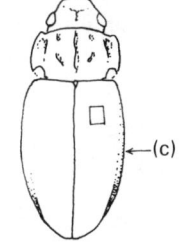

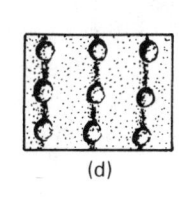

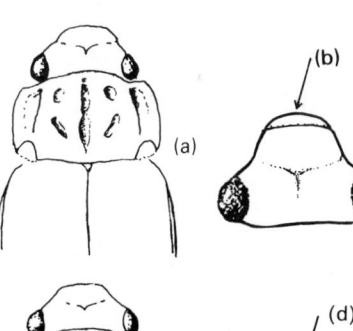

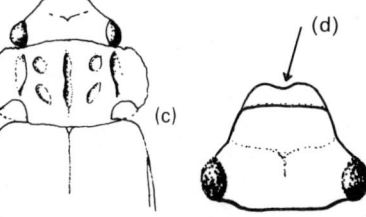

13	Central area of the pronotum (see the introduction to Key G11, p. 135) as long as wide, with fine serrations on the lateral margins visible at ×25 (a); upper surface dull; [size 1.8–2.1 mm] . . . . .
(6)	***O. subinteger*** v. ***lejolisii***

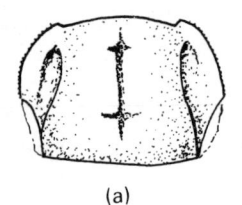

(a)

— Central area of the pronotum much wider than long (b), without serrated edges; upper surface shining 14

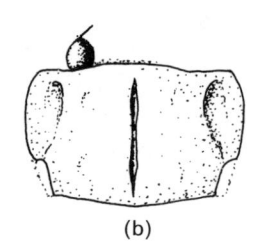

(b)

14	No grooves along the elytral puncture rows; in each row, the punctures are separated by about their own diameter; glossy bronze beetle; [size 2.0–2.3 mm] .
(13)	***O. aeneus***

— Puncture rows on elytra marked by shallow grooves; individual punctures are less than their own diameter apart and each with a tiny hair, just visible at ×80 (a); dark bronze beetle; [size 1.5–2.0 mm] . .
O. minimus

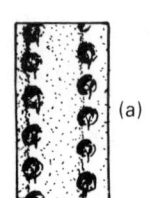

(a)

KEY H   DRYOPIDAE

Family Characters: antennae short with greatly enlarged basal segments; beetles covered with hairs; last segment of all tarsi long.
Size range 3.0–5.5 mm.
2 genera: Dryops *7 spp.*
            Helichus *1 sp.*

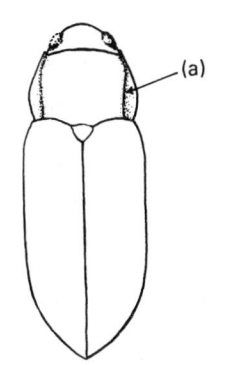

1   Pronotum with a strong longitudinal furrow on each side (a); hairs long and upstanding; [size 3.0–4.5 mm]   . ***Dryops***

   7 spp: *anglicanus* Edwards
            auriculatus (Geoffroy)
            ernesti des Gozis
            luridus (Erichson)
            nitidulus (Heer)
            similaris Bollow
            striatellus (Fairmaire & Bristout)

   The separation of these species depends on the form of the genitalia (which are figured by Olmi (1978), cleared in KOH and mounted on slides), which is beyond the intended scope of this Key.

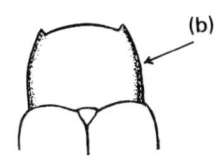

— Pronotum without furrows (b) (although a narrow rim is present); hair short; [size 4.5–5.5 mm] . ***Helichus substriatus***

TABLE 1

Colour Guide to the smaller species (less than 20 mm) of Colymbetinae and Dytiscinae (larger beetles in these subfamilies = Dytiscus) (The Family Key and Key D must be consulted before using this guide)

Elytra colour	Pronotal patterning and approximate sizes			
(a) Mottled yellow & black (use × 10)	Black bar across centre enclosing yellow bar (14–16 mm)	Black bar front and hind margin (14–16 mm)	Two black spots side by side in centre (8 mm)	Other spot and band patterns or all pale (9–12 mm)
	Acilius sulcatus canaliculatus Key D2c	*Graphoderus bilineatus cinereus zonatus* Key D2d	*Agabus nebulosus* Key D1b	*Rhantus aberratus suturellus exsoletus frontalis suturalis* Key D1a

(b) Dark with yellow spots or bands	Uniformly dark (7–8 mm)	Dark crescent hind margin (12–13 mm)	Pale band across (7–9 mm)
	Agabus didymus undulatus Key D1b	*Hydaticus transversalis* Key D2b	*Platambus maculatus* Key D1

(c) Dark with full-length wide pale margins	Pale lateral margins (10–11 mm)	Dark crescent hind margin (12–15 mm)
	Ilybius fuliginosus Key D1b	*Hydaticus seminiger transversalis* Key D2b

(Table 1 is continued opposite)

Table 1 (continued)

Elytra colour	Pronotal patterning and approximate sizes			
(d) Uniform red- or mid-brown (with, at most, faint streaks or spots)	Uniform and darker than elytra (6–10 mm)	Darker than elytra with pale borders	Same as elytra, reddish	Same as elytra, mid-brown (8–10 mm)
	Agabus congener labiatus Key D1b	(7–8 mm) *Agabus sturmii arcticus uliginosus paludosus* Key D1b	(7–8 mm) *Copelatus haemorrhoidalis* Key D1	*Agabus brunneus conspersus* Key D1b
		(16–18 mm) *Colymbetes fuscus* Key D1	(9–11 mm) *Agabus bipustulatus* Key D1b	
			(11.5 mm) *Ilybius fenestratus* Key D1b	

Elytra colour	Pronota are uniform and dark in all the species grouped here; size will separate some species; refer to Keys D1 & D1b to separate genera in the middle size range. Approximate Sizes		
(e) Uniform black or very dark (with, at most, faint streaks or spots)	c.15 mm	9–11 mm	6–8 mm
	Ilybius ater Key D1b	*Rhantus grapii Ilybius aenescens fenestratus guttiger 4-guttatus subaeneus Agabus bipustulatus biguttatus guttatus melanarius*	*Agabus affinis chalconatus labiatus melanarius melanocornis striolatus uliginosus unguicularis* Key D1b

TABLE 2
Colour Guide to Hydroporinae
This guide should be used to complement the Keys D4 to D4g. The Family Key and Key D must be consulted first.

Note: some species appear more than once where the colour pattern varies.

Are the elytra of uniform colour? . . go to . . . . Section A

or is there a pattern of:
- longitudinal stripes (with or without blotches)? . . . . Section B
- blotches, spots or bands across (no longitudinal stripes)? . . . . Section C
- paler patches on the "shoulders" only? . . . . Section D
- flecks or faint mottling on brown background? . . . . Section E

Section A—Elytra uniform (if any flecks, see Section E)
Compare colour of head (HD) and centre of pronotum (PN) with elytra:

HD &PN darker	HD and/or PN lighter	All same = brown	All same = Black or V. dark	
			Elytra hairy	Not hairy
Laccornis oblongus Key D4	*Hydroporus obscurus memnonius gyllenhalii* Key D4f	*Bidessus unistriatus* Key D4c	*Hydroporus pubescens discretus* Key D4f	*Hydroporus morio glabriusculus nigrita longulus longicornis melanarius cantabricus* Key D4f
Hydroporus neglectus elongatulus tristis umbrosus Key D4f	*Hydrovatus clypealis* Key D4	*Hyphydrus ovatus* Key D4a		
		Hydroporus scalesianus angustatus Key D4f		

Section B—Longitudinal stripes (with or without blotches)
(a) Pattern strongly marked = yellow and brown/black
(number of stripes = both elytra counted together)

Dark colour predominant	Pale colour predominant			
	Stripes run full length of elytra			No stripes on shoulder
	9 lines or more		5 lines	
Potamonectes griseostriatus depressus Key D4g	*Scarodytes halensis* Key D4	*Graptodytes flavipes* Key D4d	*Hygrotus versicolor 5-lineatus* Key D4b	*Coelambus confluens* Key D4b
	Coelambus nigrolineatus 9-lineatus parallelogrammus Key D4b	*Oreodytes alpinus davisii 7-trionalis sanmarkii* Key D4e		
	Potamonectes assimilis elegans Key D4g			

continued opposite

A Key to the Adults of British Water Beetles 145

Section B continued
(b) Pattern less contrasted = chestnut and darker brown
(number of stripes = both elytra counted together)

4 Pale lines	9 Dark lines
Graptodytes bilineatus granularis Key D4d	*Coelambus impressopunctatus* Key D4b *Porhydrus lineatus* Key D4

Section C—Blotches, spots, bands across (no longitudinal stripes)

Black with 12 yellow spots	Brown + Yellow bands across	Paler colour predominant	Darker colour predominant	
Stictotarsus 12-pustulatus Key D4	*Bidessus minutissimus* Key D4c	*Hyphydrus aubei* Key D4a	*Graptodytes pictus* *Stictonectes lepidus* Key D4d	*Hygrotus decoratus inaequalis* Key D4b *Suphrodytes dorsalis* *Hydroporus palustris* Key D4f

Section D—Paler markings on 'shoulders' only
(colours given refer to the hind half of elytra in mature specimens—check both columns)

Red or Brown	Black
Deronectes latus Key D4 *Hydroporus erythrocephalus marginatus rufifrons incognitus* Key D4f	*Hydroporus tessellatus pubescens planus palustris striola* Key D4f

Section E—Elytra brown with flecks or mottling (see also Section A)

Hyphydrus ovatus Key D4a	*Hydroglyphus pusillus* Key D4c	*Hydroporus obsoletus ferrugineus* Key D4f

Table 3. Sizes of water beetle genera.

LENGTH IN MM (front of head to tip of elytra, except Elmidae,
Dryopidae = front of pronotum to tip of elytra).

Hygrobia	HYGROBIIDAE
Gyrinus / Orectochilus	GYRINIDAE KEY A
Haliplus / Brychius, Peltodytes	HALIPLIDAE KEY B
	NOTERIDAE KEY C
Acilius / Graphoderus / Hydaticus / Dytiscus	Dytiscinae KEY D2
Colymbetes / Ilybius / Rhantus / Agabus / Platambus, Copelatus	Colymbetinae KEY D1
Laccophilus	Laccophilinae KEY D3
Stictotarsus / Hydroporus / Potamonectes / Laccornis / Hyphydrus / Deronectes / Hygrotus, Coelambus / Scarodytes, Oreodytes / Graptodytes, Stictonectes / Porhydrus / Hydrovatus / Bidessus, Guignotus	Hydroporinae KEY D4
	DYTISCIDAE KEY D
Donacia / Plateumaris / Macroplea	CHRYSOMELIDAE KEY E
Stenelmis / Macronychus / Limnius / Elmis / Riolus, Normandia, Oulimnius / Esolus	ELMIDAE KEY F
Helophorus / Sphercheus / Coelostoma / Cercyon / Hydrochus	KEYS G, G1, G2, G3
Hydrochara / Limnoxenus / Hydrobius / Enochrus / Helochares / Berosus / Cymbiodyta / Laccobius / Anacaena, Paracymus / Chaetarthria	Hydrophilus / Hydrophilinae KEYS G4, G5, G6, G7, G8
	HYDROPHILIDAE KEY G
Limnebius, Ochthebius, Hydraena	HYDRAENIDAE KEYS G9, G10, G11
Helichus / Dryops	DRYOPIDAE KEY H

SPECIES CHECKLIST, DISTRIBUTIONS & HABITAT NOTES

Names
Names and authorities are based on Pope 1977 except for species new to Britain since 1977, for which references are given as they occur in the following checklist. Synonyms found in Balfour-Browne (B-B) (1940, 1950 or 1958) and Joy (1932) are given in brackets.

Distribution and habitats
The letters in square brackets give an indication of the areas in Britain and Ireland where the beetle may be found:
  S = Scotland
  N = Northern England
  W = South-west England and Wales
  E = South-eastern England, including East Anglia
  I = Ireland.

Capital letters denote that the species may be quite likely to turn up in suitable habitats in that area; lower case letters that it may occasionally do so.

These distributions are based on information in Balfour-Browne's three volumes, updated by Dr G. N. Foster. A nationwide water beetle mapping scheme is in progress to bring our knowledge up to date. Anyone who can contribute records is asked to contact the Biological Records Centre, Institute of Terrestrial Ecology, Monks Wood Experimental Station, Abbots Ripton, Huntingdon PE17 2LS, for information and record cards. More advice can be found in the Balfour-Browne Club Newsletter or from Dr G. N. Foster (see Introduction).

The number at the end of each entry is the page reference of the couplet at which the species keys out.

Acilius Key D2c
  canaliculatus (Nicolai) [SNi] fens and pools 53
  sulcatus (L.) [SNWEI] large areas of water, ponds 53
Agabus Key D1b
  affinis (Paykull) [SNWeI] acid pools and wet moss 47
  arcticus (Paykull) [SNwi] lakes 42
  biguttatus (Olivier) [SNWei] streams and wells 45
  bipustulatus (L.) [SNWEI] wide range of habitats 39
  brunneus (Fab.) [w] shallow gravelly streams 44
  chalconatus (Panzer) [nWEI] acid water, often shaded 46
  congener (Thunberg) [SNwei] acid pools and fens 43
  conspersus (Marsham) [sNWEi] brackish pools and drains 41
  didymus (Olivier) [NWE] slow running water, lowland 40
  guttatus (Paykull) [SNWEi] springs and streams 45
  labiatus (Brahm) [snwEi] stagnant and temporary water 46
  melanarius Aubé [sNwe] shallow spring-fed pools in woods 39
  melanocornis Zimmermann [SNWEI] acid water 46
  nebulosus (Forster) [SNWEI] ponds, especially new ones 41
  paludosus (Fab.) [SNWEI] running water with vegetation 43
  striolatus (Gyllenhal) [e] fen carr 39
  sturmii (Gyllenhal) [SNWEI] stagnant water 43
  uliginosus (L.) [snwe] marshy and temporary pools 45
  undulatus (Schrank) [nwe] fen pools and drains 40
  unguicularis Thomson [SNwei] marshy pools 47
Anacaena Key G8
  bipustulata (Marsham) [nWE] streams, rivers, pits 128

globulus (Paykull) [SNWEI] running water and damp shaded ground 128
limbata (Fab.) [SNWEI] marshes and pools 127
lutescens (Stephens) [SNWEI] acid waters (Berge Henegouwen 1986) 128
Berosus Key G4
  affinis Brullé [nWE] silt ponds, drains, sometimes brackish 118
  luridus (L.) [snwEi] marsh drains and ponds 118
  signaticollis (Charpentier) [nwEi] ponds with silt 117
  spinosus (von Steven) [e] brackish water 117
Bidessus (see also *Hydroglyphus*) Key D4c
  minutissimus (Germar) [swi] running water, lakes 66
  unistriatus (Schrank) [e] dykes and meres 66
Brychius Key B
  elevatus (Panzer) [SNWEI] running water and wave-washed lakes 22
Cercyon Key G3
  bifenestratus Küster [we] damp places in sand pits 116
  convexiusculus Stephens [snwE] fen litter 116
  depressus Stephens [snwei] under decaying seaweed 114
  granarius Erichson [we] on vegetation in marsh drains 115
  littoralis (Gyllenhal) [SNWEI] under decaying seaweed 113
  marinus Thomson [snwei] damp places 116
  sternalis Sharp [we] fen litter 116
  tristis (Illiger) [snwei] bogs and fens 115
  ustulatus (Preyssler) [snwei] wet mud at edge of ponds 114
Chaetarthria Key G
  seminulum (Herbst) [snwei] moss and mud in fens and bogs 98
Coelambus (= *Hygrotus* in B-B) Key D4b
  confluens (Fab.) [sNwEi] ponds, temporary waters 63
  impressopunctatus (Schaller) [sNwEi] ponds, drain vegetation 65
  nigrolineatus Steven (= *lautus* Schaum) [e] Kent gravel pit (Carr, 1984b) 63
  novemlineatus (Stephens) [Sni] lochs 65
  parallelogrammus (Ahrens) [nwEi] brackish water 65
Coelostoma Key G
  orbiculare (Fab.) [SNWEI] damp places in fens and moss 100
Colymbetes Key D1
  fuscus (L.) [SNWEI] ponds and ditches 33
Copelatus Key D1
  haemorrhoidalis (Fab.) [snWEI] ponds and drains 34
Cymbiodyta Key G
  marginella (Fab.) [snWEi] stagnant water with vegetation 101
Deronectes (see also *Potamonectes*) Key D4
  latus (Stephens) [snwe] clear rivers and streams 59

Dytiscus Key D2a
  circumcinctus Ahrens [nwei] fens and ponds 52
  circumflexus Fab. [nwEi] brackish, sometime large inland ponds 52
  dimidiatus Bergstraesser [nwe] fens 52
  lapponicus Gyllenhal [Swi] upland lochs 52
  marginalis L. [SNWEI] ponds and drains 52
  semisulcatus Müller [SNWEI] stagnant shallow water 52
Elmis (= *Helmis* in Joy) Key F
  aenea (Müller) [SNWEI] running water in riffles 89
Enochrus Key G6
  affinis (Thunberg) [SNweI] *Sphagnum* pools 122
  bicolor (Fab.) (= *maritimus* in B-B) [nwEi] brackish water 120
  coarctatus (Gredler) [sNWEI] pools with rich vegetation 122
  fuscipennis (Thomson) [snWeI] bogs (Hansen, 1987) 122
  halophilus (Bedel) [wei] brackish water (Hansen, 1987) 121
  isotae Hebauer [we] fens (Foster, 1984) 122
  melanocephalus (Olivier) [nWEi] brackish water 120
  ochropterus (Marsham) [snwei] detritus pools and drains 121
  quadripunctatus (Herbst) [se] freshwater 122
  testaceus (Fab.) [snWEI] dykes and ponds 121
Esolus Key F
  parallelepipedus (Müller) [SNWeI] running water, mud at edge 89
Georissus Key G
  crenulatus (Rossi) [snwe] crumbly damp silt by running water, sand pits 95
Graphoderus Key D2d
  bilineatus (Degeer) [e ?extinct] 54
  cinereus (L.) [we] stagnant water 54
  zonatus (Hoppe) [w/e] one bog site 54
Graptodytes (= *Hydroporus* in B-B) Key D4d
  bilineatus (Sturm) [nwei] coastal pools, Essex: sometimes inland 68
  flavipes (Olivier) [we] heath pools 67
  granularis (L.) [snwEi] swampy areas of ponds, fens, bogs 68
  pictus (Fab.) [sNWEi] slow-flowing drains and ponds 67
Guignotus (= *Bidessus* in B-B, Joy)—see *Hydroglyphus*
Gyrinus Key A
  aeratus Stephens [SnweI] lakes, canals, slow reaches of rivers 19
  bicolor Fab. (= *paykulli* Ochs.) [snwei] reeds at edge of open water 20
  caspius Ménétriés [SNWEI] fen drains and pools, mainly coastal 20
  distinctus Aubé (= *colymbus* in B-B) [snei] lakes and drains 21
  marinus Gyllenhal [sNWEI] fresh, peaty and brackish water 19

minutus Fab. [SnwI] lakes, ponds and drains 18
natator (L.) [I] acid lakes 21
opacus Sahlberg [s] upland small pools (lochans) 19
substriatus Stephens (=*natator* in B-B, Joy) [SNWEI] fresh, peaty, sometimes brackish water 21
suffriani Scriba [we] edges of pools 20
urinator Illiger [nWei] running water, lowland 18

Haliplus Key B1
apicalis Thomson [sNwEi] brackish water 28
confinis Stephens [SNWEI] fen ditches and dykes, pools and streams 24
flavicollis Sturm [SNWEI] running water and lakes 25
fluviatilis Aubé [SNWEI] rivers, sometimes lakes and pools 28
fulvus (Fab.) [SNWEI] large areas of water or bog pools 26
furcatus Seidl. [we] large ponds 28
heydeni Wehncke [nWE] small grassy ponds and ditches 28
immaculatus Gerhardt [sNWEi] canals, lakes, clean ponds 27
laminatus (Schaller) [nE] canals, rivers and silt ponds 25
lineatocollis (Marsham) [SNWEI] mainly slow-running water 23
lineolatus Mannerheim [SNWEI] rivers and lakes 28
mucronatus Stephens [nwe] clay and gravel pits 26
obliquus (Fab.) [sNwEI] pools, especially with *Chara* 24
ruficollis (Degeer) [SNWEI] ponds and ditches 27
variegatus Sturm [wei] fen pools 26
varius Nicolai [e] ponds, E. Sussex (Parry 1983) 23
wehnckei Gerhardt [SNWEI] ponds, canals, rivers 27

Helichus Key H
substriatus (Müller) [nwe] under submerged logs and stones 141

Helmis—see *Elmis*

Helochares Key G5
lividus (Forster) [nwE] ponds 119
obscurus (Müller) [we] old fen (Hansen, 1987) 119
punctatus Sharp [snWEI] peat and *Sphagnum* pools 119

Helophorus Key G1
aequalis Thomson [SNWEI] grassy pools and ditches (= *aquaticus* L in B-B) 104
alternans Gené [snwEi] weedy ponds, mostly coastal 103
arvernicus Mulsant [SNWei] sides of rivers 105
brevipalpis Bedel [SNWEI] ubiquitous, temporary waters 105
dorsalis (Marsh.) [nwE] mainly shaded pools 108
flavipes Fab. [SNWEI] acid water, *Sphagnum* pools 109
fulgidicollis Motsch. [snWEi] saltmarshes 108
grandis Illiger [SNWEI] grassy pools and ditches 104
granularis (L.) [snwei] grassy ponds 108
griseus Herbst [nwE] grassy ponds and fens 109
laticollis Thomson [w] heath pools in New Forest 107
longitarsis Woll. [we] silt ponds and temporary waters 109
minutus Fab. [SNWEI] grassy ponds 109
nanus Sturm [nwEi] acid water and fens 106
obscurus Mulsant [SNWEI] ponds 109
strigifrons Thomson (=*flavipes* in B-B) [snwei] temporary waters with rushes and sedges 107
tuberculatus Gyllenhal [snwe] peat moss, following burning 101

Hydaticus Key D2b
seminiger (Degeer) [wEi] pools, often shaded 53
transversalis (Pontoppidan) [we] dykes and pits 53

Hydraena Key G10
britteni Joy [SNwei] in fens, grassy streams in south 134, 135
gracilis Germar [SNWEI] running water 132
minutissima Stephens [snwi] gravel, moss-covered stones 133
nigrita Germar [snwei] gravel and stones in rivers, often shaded 134
palustris Erichson [e] fen 133
pulchella Germar [snwei] streams with muddy backwaters 133
pygmaea Waterhouse [snwi] in moss, small streams 132
riparia Kugelann [SNWEI] ditches and pools 134, 135
rufipes Curtis [snwei] rivers, mossy streams 134
testacea Curtis [snWEi] stagnant water or muddy streams 131

Hydrobius Key G
fuscipes (L.) [SNWEI] detritus pools 98

Hydrochara Key G
caraboides (L.) [w] ponds and ditches on Somerset levels 93

Hydrochus Key G2
angustatus Germar [nWEi] muddy water, mainly on heaths 113
brevis (Herbst) [sei] swamps 112
carinatus Germar [e] fen 112
elongatus (Schaller) [wei] ponds and drains, often among reeds 112
ignicollis Motschulsky [we] marshes and fen 112
nitidicollis Mulsant (=*interruptus* in B-B) [w] muddy areas in ponds and streams 113

Hydroglyphus (= *Guignotus* in Pope, *Bidessus* in B-B & Joy) Key D4c
pusillus (Fab.) (=*geminus* in B-B, Joy) [nwe] new ponds, heath pools, mossy ditches 66

Hydrophilus Key G
piceus (L.) [we] dykes, marshes 93

Hydroporus (see also *Graptodytes, Porhydrus, Stictonectes,* and *Suphrodytes*) Key D4f
  angustatus Sturm [SNWEI] fens and ponds 74, 76
  cantabricus Sharp [w] heath pools 78
  discretus Fairmaire [SNwEi] muddy springs, streams 73
  dorsalis (Fab.)—see *Suphrodytes*
  elongatulus Sturm [se] fen mosses 81
  erythrocephalus (L.) [SNWEI] marsh, peat mosses in pools 79
  ferrugineus Stephens [SNwe] springs and pools 75
  glabriusculus Aubé [sei] shallow fens 77, 80
  gyllenhalii Schiödte [SNWEI] acid water 76
  incognitus Sharp [SNWEi] peaty water, in woods 82
  longicornis Sharp [snwei] spring-fed bogs, fens 78
  longulus Mulsant [SNwei] springs, peaty streams 78
  marginatus (Duftschmid) [we] stagnant or running water on chalk or limestone 72
  melanarius Sturm [SNwei] peat mosses 78
  memnonius Nicolai [snwei] matt female [nwe] shining female [si]; stagnant water with dead leaves 76
  morio Aubé [SNwei] bogs, mainly upland 77
  neglectus Schaum [nwE] shaded, acid water 80
  nigrita (Fab.) [SNWEI] ponds, ditches, *Sphagnum* 78
  obscurus Sturm [SNWEI] acid water, *Sphagnum* 76
  obsoletus Aubé [snwi] springs and small streams 75
  palustris (L.) [SNWEI] ponds and slow water 81
  planus (Fab.) [SNWEI] lowland pools, often temporary 73
  pubescens (Gyllenhal) [SNWEI] peaty, fresh or brackish water 73
  rufifrons (Müller) [snwe] fens, river oxbows 79
  scalesianus Stephens [nei] boggy ponds 75
  striola (Gyllenhal) [SNWEI] pools, ditches and marshes 82
  tessellatus Drapiez [sNWEI] pools, running water, saltmarshes 72
  tristis (Paykull) [SNWei] fen marshes 81
  umbrosus (Gyllenhal) [SNWeI] fen marshes 81
Hydrovatus Key D4
  clypealis Sharp [we] ponds, coastal 57
Hygrobia Family Key
  hermanni (Fab.) [nWEI] silt and detritus ponds 15
Hygrotus (see also *Coelambus*) Key D4b
  decoratus (Gyllenhal) [nwE] ponds and drains 64
  inaequalis (Fab.) [SNWEI] ponds, bays of lakes, slow water 64
  quinquelineatus (Zett.) [SnI] lochs, fen drains 64
  versicolor (Schaller) [nwEi] ponds, drains, slow rivers 64
Hyphydrus Key D4a
  aubei Ganglbauer [Channel Islands] 62
  ovatus (L.) [sNWEI] still or slow-running water with vegetation 62
Ilybius Key D1b
  aenescens Thomson [snwei] acid water, deeply-flooded *Sphagnum* 48
  ater (Degeer) [SNWEI] stagnant water 38
  fenestratus (Fab.) [snwE] ponds, pits, rivers, marshes 47
  fuliginosus (Fab.) [SNWEI] wide range of habitats 40
  guttiger (Gyllenhal) [sNwEi] stagnant water, fens and bogs 48
  quadriguttatus (Lacordaire and Boisduval) (=*obscurus* in B-B) [nWEI] stagnant water 48
  subaeneus Erichson [sNE] detritus ponds 47
Laccobius Key G7
  atratus (Rottenberg) [snWei] acid water, in flushes 124
  atrocephalus Reitter [SnWei] mud, silt at edge of streams 125
  biguttatus Gerhardt [snwEi] silt ponds, coastal drains 124
  bipunctatus (Fab.) (=*alutaceus* in B-B) [SNWEI] mud 125
  minutus (L.) [SNWEI] ponds and lakes 124
  obscuratus Rottenberg, one record, "London" (Gentili, 1977), running water 123
  simulatrix d'Orchymont [e] one gravel pit site, E. Sussex (Foster & Phillips, 1983) 124
  sinuatus Motschulsky [nwe] slow-flowing drains and new ponds 126
  striatulus (Fab.) [SNWEI] slow trickles, streams, riverbank mud 123, 126
Laccophilus Key D3
  hyalinus (Degeer) [nWEi] slow-running water 55
  minutus (L.) [sNWEI] ponds and ditches, lowland 55
  obsoletus Westhoff (=*variegatus* (Germar)) [ne] coastal ponds and weedy ditches 55
Laccornis Key D4
  oblongus (Stephens) [snwei] mossy pools 59
Limnebius Key G9
  aluta (Bedel) [nwei] fens, in mud 130
  crinifer (Rey) [e] one acid pool site, Kent (Carr, 1984a) 130
  nitidus (Marsham) [snWEi] marsh drains, muddy streams and ponds 130
  papposus Mulsant [nwE] fen 129
  truncatellus (Thunberg) [SNWEI] running water 129, 130
Limnichus Family Key
  pygmaeus (Sturm) [we] beside running water 17
Limnius (see also *Oulimnius*) Key F
  volckmari (Panzer) [SNWEI] running water, highland lochs 89
Limnoxenus Key G
  niger (Zschach) [wE] fen drains 98
Macronychus Key F
  quadrituberculatus Müller [w] on logs in rivers 88
Megasternum Key G
  obscurum (Marsham) 97
Normandia Key F1
  nitens (Müller) [nw] running water 90

A Key to the Adults of British Water Beetles 151

Noterus Key C
 clavicornis (Degeer) (= *capricornis* in B-B) [sNWEI] still water, often in weed rafts 30
 crassicornis (Müller) (= *clavicornis* in B-B) [snwEI] still water, in fens 30

Ochthebius Key G11
 aeneus Stephens [e ?extinct] heath pools, brackish water 140
 auriculatus Rey [snwei] brackish water, in grass 139
 bicolon Germar [snwei] mud by running water 139
 dilatatus Stephens (= *impressicollis* in B-B) [SNWEI] muddy water 139
 exaratus Mulsant [wE] mud by coastal ponds 136
 exsculptus Germar [SnWei] shallow running water 138
 lenensis Poppius [s] brackish pools 137
 marinus (Paykull) [sNWEi] brackish pools 137
 minimus (Fab.) [SNWEI] stagnant water, ponds, usually in mud 140
 nanus Stephens [wEi] marsh drains, coastal ponds 139
 poweri Rye (= *metallescens* in B-B) [wi] wet cliffs 139
 punctatus Stephens [snwei] brackish water 138
 pusillus Stephens [we] ponds, marshes 136
 subinteger s. lejolisii Mulsant & Rey [SnWeI] rock pools above high water 140
 viridis Peyron [snwEi] brackish and heath pools 137

Orectochilus Key A
 villosus (Müller) [SNWeI] running water, wave-washed lake edges 18

Oreodytes (see also *Scarodytes*) Key D4e
 alpinus (Paykull) [s] running water (Foster & Spirit, 1986) 68
 davisii (Curtis) (= *borealis* in B-B) [SNwi] rocky streams 69
 sanmarkii (Sahlberg) (= *rivalis* in B-B) [SNWei] running water and lakes 69
 septentrionalis (Sahlberg) [SNWI] running water and lakes 69

Oulimnius (= *Limnius* in Joy) Key F2
 major (Rey) [we] running water and fen drains (Parry, 1980) 92
 rivularis (Rosenhauer) [e] fen drains 91
 troglodytes (Gyllenhal) [snWe] running water and lakes 91
 tuberculatus (Müller) [SNWEI] running water and lakes 92

Paracymus Key G8
 aeneus (Germar) [Isle of Wight] brackish water 128
 scutellaris (Rosenhauer) [snWei] shallow acid water 128

Peltodytes Key B
 caesus (Duftschmid) [wE] fenland drains and quarry ponds 22

Platambus Key D1
 maculatus (L.) [SNWE] running water and wave-washed lakes 34

Porhydrus (= *Hydroporus* in B-B) Key D4
 lineatus (Fab.) [snwEI] muddy ponds and ditches, lowland 61

Potamonectes (= *Deronectes* in B-B) Key D4g
 assimilis (Paykull) [SNwei] lakes, ponds 84
 depressus s. depressus (Fab.) [SnI] lochs and rivers 85
 depressus s. elegans (Panzer) [SNWE] rivers and lochs 85
 griseostriatus (Degeer) [Snwi] hill lochs with peat over gravel 84

Rhantus Key D1a
 aberratus Gemminger & von Harold (= *adspersus* in B-B) [e ?extinct] fen 37
 exsoletus (Forster) [SNwEI] ponds and drains, edges of lakes 37
 frontalis (Marsham) (= *notatus* in B-B) [swEi] sandy pools, subsidence ponds on peat in Scotland 36
 grapii (Gyllenhal) [nwEi] ponds and fen drains 35
 suturalis (MacLeay) (= *pulverosus* in B-B) [nwE] silt and detritus pools 36
 suturellus Harris (= *bistriatus* (Bergstraesser)) [SNwei] peat pools 37

Riolus Key F1
 cupreus (Müller) [snwe] base-rich running water and lakes 90
 subviolaceus (Müller) [snwe] base-rich running water 90

Scarodytes (= *Oreodytes* in B-B) Key D4
 halensis (Fab.) [ne] slow-flowing streams and silt ponds 62, 69

Spercheus Key G
 emarginatus (Schaller) [e ?extinct] swamps, muddy water 94

Stenelmis Key F
 canaliculata (Gyllenhal) [nwe] deep water in rivers and a lake 88

Stictonectes (= *Hydroporus* in B-B) Key D4d
 lepidus (Olivier) [sNWeI] streams, dam ponds, quarry pools 67

Stictotarsus Key D4
 duodecimpustulatus (Fab.) [SNWEI] running water, lakes 56

Suphrodytes (= *Hydroporus*) Key D4f
 dorsalis (Fab.) [NwEi] ponds, often shaded 71

AIDGAP PUBLICATIONS

The Field Studies Council (FSC) will have published thirteen AIDGAP keys by the end of 1988:—

Hiscock, Sue (1979). *A field key to the British brown seaweeds* (FSC Publication 125).

Sykes, J. B. (1981). *An illustrated guide to the diatoms of British coastal plankton* (FSC Publication 140).

Unwin, D. M. (1981). *A key to families of British Diptera* (FSC Publication 143).

Crothers, John & Marilyn (1983). *A key to the crabs and crab-like animals of British inshore waters* (FSC Publication 155).

Cameron, R. A. D., Eversham, B. & Jackson, N. (1983). *A field guide to the slugs of the British Isles* (FSC Publication 156).

Unwin, D. M. (1984). *A key to the families of British Coleoptera and Strepsiptera* (FSC Publication 166).

Willmer, Pat (1985). *Bees, ants and wasps—the British Aculeates* (FSC Occasional Publication 7).

Pankhurst, R. J. and Allinson, J. (1985). *British Grasses: a punched-card key to grasses in the vegetative state* (FSC Occasional Publication 10).

King, P. (1986). *Sea Spiders: a revised key to the adults of littoral pycnogonida of the British Isles* (FSC Publication 180).

Croft, P. (1986). *A key to the major groups of British Freshwater Invertebrates* (FSC Publication 181).

Hiscock, Sue. (1986). *A field guide to the British Red Seaweeds* (FSC Occasional Publication 13).

Tilling, S. M. (1987). *A key to the major groups of British Terrestrial Invertebrates* (FSC Publication 187).

Friday, L. E. (1988). *A key to the adults of British water beetles* (FSC Publication 189).

Another key, published before the AIDGAP project was initiated, has been fully tested and revised:—

Haslam, S. M., Sinker, C. A. & Wolseley, P. A. (1975). *British Water Plants* (FSC Publication 107).

These, and many other titles, may be purchased when visiting Field Studies Council Centres or may be ordered through the post from:—

"Field Studies", Nettlecombe Court, Williton, Taunton, Somerset TA4 4HT,
or from
The Richmond Publishing Company Ltd., Orchard Road, Richmond, Surrey TW9 4PD

A complete list of titles and prices is available from either of these addresses.